Sustainable Livestock Production

Sustainable Livestock Production

Dean Campbell

Editor

KOROS PRESS LIMITED

London, UK

Sustainable Livestock Production

© 2012

Printed in 2017 for Sale in the Indian Subcontinent

Published by
Koros Press Limited
3 The Pines, Rubery B45 9FF, Rednal,
Birmingham, United Kingdom

Tel.: +44-7826-930152
Email: info@korospress.com
www.korospress.com

ISBN: 978-1-78163-130-0

Editor: Dean Campbell

Printed in UK

10 9 8 7 6 5 4 3 2 1

British Library Cataloguing in Publication Data
A CIP record for this book is available from the British Library

Exclusively distributed by CBS Publishers & Distributors Pvt. Ltd.
Sales & Distribution Rights only for India, Pakistan, Bangladesh, Sri
Lanka, Nepal and Bhutan.This book is not to be sold outside these
territories.

Contents

Preface

Since farm animal breeding is a commercial activity, it might be suggested that breeders would be best advised to focus on business imperatives and need not detain themselves with questions about an issue like sustainability. Such questions, it might be added, can be left to politicians, officials and representatives of interest groups — that is, to those working in the formulation, implementation and maintenance of agricultural policy. This attitude is understandable. However, we suggest that farm animal breeders have several good reasons to get involved in discussions about sustainability. First, there is no avoiding the fact that, in Europe and throughout the developed world, the concept of sustainability has come to occupy a prominent role in much present-day planning of the future of agricultural production as a whole. Clearly, farm animal breeders need to engage with this planning activity, but there is little chance of their doing this, still less of their influencing planning decisions, if they do not reflect carefully on the implications of sustainability for the breeding sector. Second, twenty-first century consumers in the developed, food-rich world appear to have a real interest in foods produced in a sustainable way. Quite what features sustainably produced foods must have is debatable.

It is also debatable what features consumers take sustainably produced foods to have. But there is no doubt that for many western consumers the word 'sustainable', like the more specific 'biodegradable' and 'recyclable', is positive enough to support a viable and growing niche-market in foods. Breeders would be unwise to ignore this social development. Third, farm animal breeders are in effect already looking at sustainability-related traits. Functional traits affecting the health and welfare of future farm animals are increasingly important, as is the preservation of genetic resources for future use, and animal health and welfare, and the preservation of genetic resources, are elements of sustainability as it is understood today. The great advantage of putting sustainability in the foreground, as we explain at the end of the next section, is that it obliges decision-makers to combine elements like these in a unified perspective. The idea of sustainability has in fact evolved over hundreds of years and has come to encompass a growing number of concerns. In its earliest deployment, it was connected with the simple aim of maintaining renewable resources for harvest and consumption in perpetuity, i.e. sustained yield. This aim was first described systematically more than 350 years ago in connection with German forestry and mining. Over the next 200 years, the concept of sustained yield became, in different interpretations, something of a lodestar — in forestry especially, but also in fisheries and other primary industries involving the regular harvesting or extraction of a natural resource.

During the twentieth-century, the goals of sustainability came to include more than that of maintaining consumable resources such as fish, firewood and fodder. In the 'Stockholm Declaration on the Human Environment' resulting from the 1972 United Nations Conference on the Human Environment, and in the 'World Conservation Strategy' released in 1980 by the World Conservation Union (IUCN), sustainability was elaborated, beyond an anthropocentric concern with human livelihood, to cover the preservation of species and ecosystems. The seminal 1987 Brundtland commission report of 1987 shifted the focus to distribution, and in particular distribution across generations, the idea being that there should be no shortterm privileges for the present generation. In policy circles, however, they had come to be regarded as indispensable components of sustainable development. In essence, a process of conceptual erosion was underway. Such erosion occurs when aims or objectives widely considered worthy of promotion are collected under a single heading in a way that allows the heading to be applied to a range of issues in various disciplines and sectors. Initially this can be rhetorically effective. The obvious problem, however, is that eroded concepts become harder and harder to deploy meaningfully as they incorporate a growing number of aims and aspirations. In particular, and most strikingly, it often becomes impossible to pursue the many ideals that fall under an eroded concept simultaneously.

In essence the book could be of vital significance for sustainable animal production and feeding technology.

—Editor

Chapter 1

Introduction

Sustainable Animal Production

Farming is a complex, multicomponent, interactive process that is dependent on land, animal, human and water resources as well as capital investment. Throughout the developing world it is practiced in many different ways and environments and with differing degrees of intensity and biological efficiency.

Animals play an integral role in many of these farming systems. Unlike the specialised and intensified livestock systems in the developed world, animal production in developing countries utilises the full range of animal outputs, many of which are returned as essential inputs to the farm production system. Manure, for example, is an important - and often the only - source of plant nutrients and, unlike chemical fertilizers, provides the organic matter necessary for maintaining soil structure, which is an important factor in erosion control.

Animal traction is another major farm input, especially in Asia, and there are an estimated 300 million draught animals of various species in the developing world. Land preparation immediately comes to mind when considering animal traction but the important contribution of animals in transporting both goods and humans should not be underestimated. Livestock also provide income on a regular basis, such as from the sale of milk, and serve as a strategic source of cash that may be drawn on as required, since animal ownership is also a rational means of storing and accumulating wealth. This function has important implications in that it finances other essential farm inputs, such as seeds and agrochemicals, while providing cash during the critical periods of the year and thereby alleviating the poverty trap as well as increasing household food security.

Food production in many developing countries is insufficient for current requirements and the situation will be further exacerbated by rapidly expanding human populations. Countries in this position cannot allow food production to stagnate. Nor can they afford, in the longer term, to produce food at the expense of the environment or their natural resource base. There is clearly a need for more sustainable increases in agricultural output in order to balance the adverse effects of intensive production and the exploitation of the resource base while, at the same time, optimising food production. Livestock serves as a means for recycling nutrients and as a source of energy and value-added production. Their complementary role within the farming system is unique and needs to be fully exploited.

However, there is a growing consensus among politicians, planners and scientists that livestock production in the developing countries is not expanding at a sufficient pace to meet the needs of the increasing world population. Growth in the output of animal products from these countries has largely been the result of increased animal populations rather than increased productivity. While the higher outputs are welcome, they do not necessarily equate with sustainable productive growth. On the contrary, in many cases they may lead to the lowering of productivity and degradation of the natural resource base. Despite the many development projects implemented over the years by national, bilateral and multilateral agencies - often with substantial capital investment - the sobering reality is that there has been little change in either the efficiency or sustainability of animal production in the developing world.

Given the increasing concern regarding conservation of the natural resource base and protection of the global environment, FAO has attached the highest priority to the sustainable development of both plant and animal agriculture. One of the initiatives undertaken by the Organisation was to convene an expert consultation in December 1990 to examine the issues relating to sustainable animal agriculture in developing countries. It became evident from this consultation that strategies to achieve sustainable livestock production would need to address matters such as the conservation of the resource base, the minimising of waste and the maximising of nutrient-recycling as well as production per unit of resource.

Technologies clearly need to be designed, taking full account of the multifaceted role of livestock in agriculture. For such technologies to be acceptable to the producer, they must be appropriate, affordable and give tangible benefits. There are already promising technologies available which optimise local feed resources, particularly crop residues, agro-industrial by-products and productive, multipurpose crops such as sugar cane and cassava.

The optimal use of the many different types of legume and their incorporation into existing farming practices will be a crucial factor in achieving sustainable agriculture. Two important animal products, manure and draught power, are grossly under utilised at present and specific programmes will be required to maximise their use.

If sustainable livestock development is to be attained, it will be necessary to assist national governments in preparing appropriate strategies and the necessary policy framework. FAO's Animal Health and Production Division has taken the initiative, incorporating strategies for sustainable livestock development in its 1992-1993 programme.

Furthermore, as a result of the expert consultation a number of case-studies have been instigated, covering the various agro-ecological zones and major production systems as a basis for determining possible criteria and parameters pertaining to sustainable animal agriculture.

Sustainable Intensive Livestock Systems for the Humid Tropics

Agriculture at the Crossroads

Since the inception of international aid, the goal of development projects in agriculture has been to increase productivity. Only now is it being realised that the production systems shaped by this narrow objective are not sustainable.

The drawbacks are many and complex: high costs, contamination of the environment, soil erosion and animal and human stress are the consequences of modern agricultural practices. Many of these are the result of the intensification process per se.

Thus, the increasing emissions of methane, perhaps the most damaging of the greenhouse gases, can be traced back to intensive rice culture and the expanding population of ruminant animals, especially in developing countries. It also appears that the effectiveness of important methane sinks, which are present in natural soil-based ecosystems, has been reduced by the burning of crop residues and heavy application of synthetic chemical fertilizers.

Chemical contamination of water and soil is a consequence of the increased use of agrochemicals in cropping. Soil erosion in both arid and humid tropical zones is largely the result of overstocking with grazing animals. Deforestation in the Amazon and in other tropical regions of Latin America is closely linked to the expansion of cattle ranching (Murgueitio, 1990).

The industrialised countries' growing concern for animal welfare is partly a reaction to the stress caused by the intensification of housing and resource management. Consumer preference for "naturally" produced food can be partially interpreted as an expression of dissatisfaction with production systems that use an excess of additives, such as antibiotics and hormones in animals and chemicals in crop production.

The pressures to liberalise world trade predicate profound changes in agricultural production systems in industrialised countries as subsidies and tariffs are gradually withdrawn. The need to restrain the use of fossil fuels in order to combat global warming will force oil prices up which, in turn, will encourage the practice of organic agriculture and add value to biomass grown for fuel and chemical substrate.

These trends add up to an impending major crisis for agriculture in the industrialised countries. They also create a unique opportunity for the tropical regions of developing countries to capitalise on the comparative advantages inherent in their rural-based economies, including their capacity to produce year-round high yields of biomass for conversion into fuel, food and feed.

Figure 1: Harvesting sugar cane for livestock

Figure 2: Sugar cane associated with tree crops on a family farm

Figure 3: Pigs being fed sugar cane juice on a small family farm

The Way Ahead

In the past, livestock production schemes in tropical developing countries were characterised more by failures than successes, largely because they attempted to transfer inappropriate (industrial) technologies, requiring expensive and often imported inputs, instead of exploiting locally available resources.

Recognition of these past errors and appreciation of the new scenarios offered by changing world climates, in both biological and economic terms, provide the rationale (Preston, 1990a) behind the hypothesis that future agricultural production systems in the tropics must be based on the following two principles:

- exploiting local comparative advantages in order to produce biomass competitively and transform it into food, feed and fuel for local consumption and sale on world markets;

- ensuring that the systems selected are economically, ecologically, sociologically and ethologically sustainable.

Biomass for Food, Feed and Fuel

The identification of high-yielding sustainable ecosystems must be the first step in any attempt to design new interventions. The products of such ecosystems must be able to serve as the principal inputs for integrated activities aimed at furnishing food, feed and fuel for immediate sale or consumption, while the by-products and residues should serve as inputs for livestock husbandry.

This activity will then contribute to earnings by providing milk and meat, farm power and, from the recycled manure, fuel and fertilizer.

Sugar Cane, Trees and Water Plants for Sustainable Livestock Production in the Tropics

The basic technology described in this article uses sugar cane, multipurpose trees and water plants as sources of biomass to provide feed for a range of livestock species as well as fuel for the farm and household.

The preferred animal species for this technology are pigs and ducks, as they readily adapt to the "unconventional" high-moisture feed resources (mainly cane juice and water plants) and have a high meat/methane production ratio.

The data show that the most productive ecosystems are perennial crops and trees grown in the tropics. From this conclusion it is a short step to the thesis that "Sugar cane and fodder trees are the logical alternatives to cereal grains as the basis of intensive livestock production and renewable energy substrate" (Preston, 1990b).

The farming system developed from these concepts in the Cauca Valley in Colombia supports extremely high levels of livestock production (in the order of 3 000 kg meat/ha/year) derived from environmentally protective perennial crops (sugar cane, nitrogen-fixing trees and water plants).

1. Perennial crops and forests in the tropics are the most productive ecosystems
2. Integrated mixed farming system based-on sugar cane, multipurpose trees and the recycling of wastes through biodigestors, ponds and earthworms.

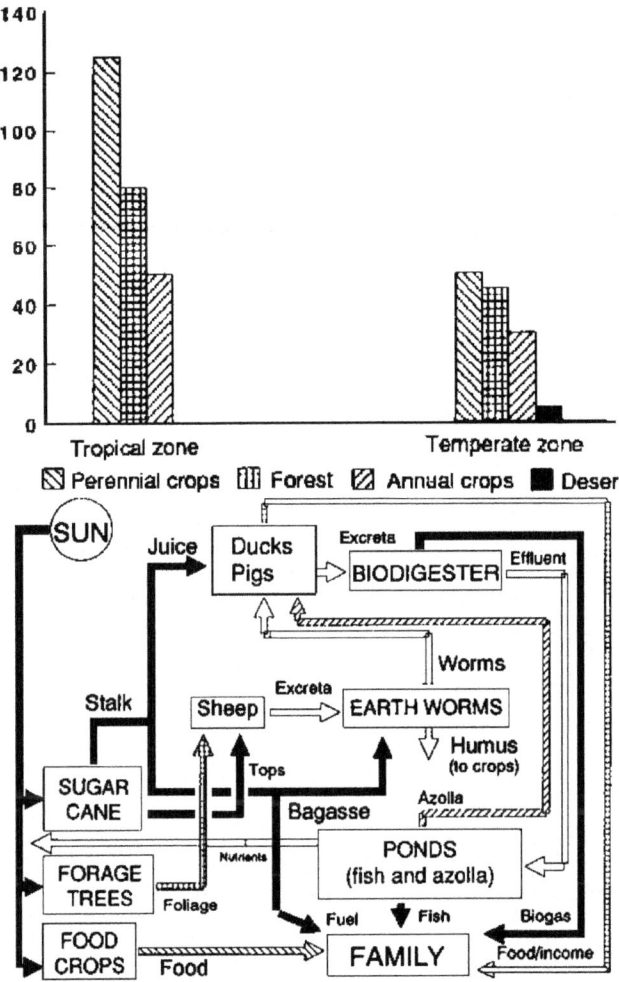

1. Performance of Pekin ducks fed sugar cane juice compared with a cereal-based (rice) diet

Performance	Cereal diet	Cane juice diet
Liveweight (kg)		
- Initial	0.727	0.72
- Final	2.75	2.51
Daily liveweight gain (g/day)	52.3	46.0
Feed intake (kg)		
- Concentrate	6.48	-
- Supplement	-	3.2
- Cane juice	-	17.4

Source: Bui Xuan Men and Vuong Van Su, 1992.

2. Productive and reproductive parameters of the flock of African hair sheep, December 1988-March 1991.

Performance	Mean value	SD[3]	n[4]
Liveweight (*kg*)			
- Birth	2.32	0.52	167
- Weaning	14.90	2.62	84
Weight gain to weaning (*g/day*)	106	33.6	84
Age at weaning (*days*)	129	45.5	84
Lambing interval (*days*)	284	85.3	44
Litter size[1]	1.16		144
Lambs born/ewe/year[2]	1.49		44
Mortality (*% all births*)			
- Perinatal	5.5		
- Birth to weaning	10.4		

Number of lambs born per parturition.

Mean number of lambs born per ewe per year.

SD = standard deviation.

n = sample number.

Source: Mejía *et al.*, 1991.

3. Mean values for the feed intake of a flock of tropical hair sheep, 1 July-31 December 1990

Diet components	Fresh basis	Dry basis
Feed intake (*kg/day*)		
- *Gliricidia* sp.	0.777	9.3[1]
- Sugar cane tops	5.640	72.4[1]
- Multinutritional block	0.121	6.2[1]
- Poultry litter	0.204	10.6[1]
- Rice polishings	0.021	1.0[1]
Total dry matter intake (*kg/day*)[2]		1.735
Total dry matter intake (*% of liveweight*)		4.493

Percentage of the total diet (dry matter basis).

For a sheep unit (on average: 1 ewe of 25 kg and 1 lamb of 14 kg).

Source: Mejía *et al.*, 1991.

3. Sugar cane juice supports high levels of performance in fattening pigs

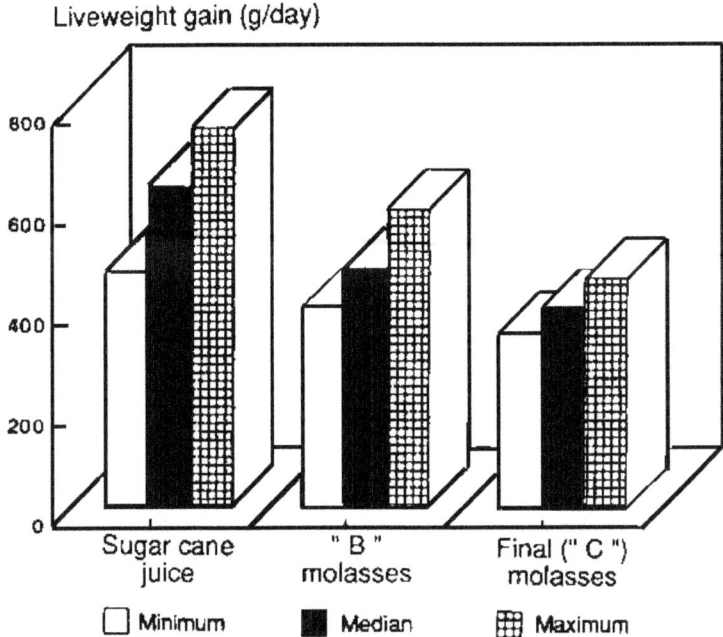

Liveweight gain (g/day)

Sugar cane juice | " B " molasses | Final (" C ") molasses

☐ Minimum   ■ Median   ▦ Maximum

Source: Sarria, Solano and Preston, 1990. Data from farm trials in Colombia.

The sugar cane stalk, after removal of the tops, is fractionated into juice and bagasse, using a simple animal-powered three-roll mill. The tree foliage is separated into leaves and twigs. The cane juice is a complete replacement for cereal grains and is the basis (75 percent) of a high-quality diet for pigs and ducks. The cane tops are fed to tropical hair sheep. Tree leaves and aquatic plants provide protein for both pigs and sheep (Becerra *et al.*, 1990; Mejía *et al.*, 1991). The bagasse and twigs are used for fuel.

The pigs and sheep are confined and their excrete recycled through plastic-bag biogas digesters, ponds (for aquatic plants) and earthworms. These downstream elements complement the system, providing additional household fuel, protein for the livestock and organic fertilizer and humus for the crops. Moreover, soil erosion, which is a serious problem in tropical grazing systems, is avoided.

The Biomass Subsystem

Sugar cane varieties, chosen for high biomass yield, are planted at twice the normal density in row widths of 75 cm, *vis-a-vis* industrial sugar production. The stalks and tops are harvested for animal feed at 12 month intervals. The dead leaves (trash) are left on the soil as mulch. The interface between the decaying mulch and the soil is where a symbiotic combination of bacteria and fungi (Patriquin, 1982) fix up to 100 kg N/ha/year. It should be noted that, in most countries, in industrial sugar production both the tops and trash are burned to facilitate harvesting. Apart from the waste of a valuable resource, this practice pollutes the environment.

At least three multipurpose tree species are now being used commercially to produce feed protein (leaves) and fuel (branches): *Gliricidia septum, Trichantera gigantea* and *Erythrina fusca*. The two former species adapt to a wide range of soil types and elevations (up to 1 800 m above sea level). The niche for which these species are not suitable (heavy clay soils with a high water-table) is the ideal habitat for *Erythrina fusca*. Leucaena is not recommended because of its high cost of harvesting in cut-and-carry systems. Both *Gliricidia* sp. and *Trichantera* sp. are planted at densities of 20 000 plants/ha, the former from seed and the latter from cuttings. A lower density (1 000 plants/ha) is recommended for *Erythrina* sp. which can be established from seed or stakes.

Two ha of sugar cane are estimated to have an annual yield of 240 tonnes of stalk which, fractionated in a three-roll mill, produces 120 tonnes of juice and 120 tonnes of bagasse (816 MJ of fuel energy). *G. septum* and/or *T. gigantea*, planted on 0.14 ha, yield an annual 8.2 tonnes of edible foliage which is fed to the sheep.

The Pig and Duck Subsystem

The technologies for feeding pigs sugar cane juice were developed in Mexico (Mena, 1981), the Dominican Republic (Fermin *et al.*, 1984) and Colombia. The feeding system for ducks was developed in Viet Nam. One pig fattened from 25 to 90 kg of liveweight consumes 1 200 kg of cane juice, 53 kg of whole soybean grain and 560 kg of fresh azolla water fern (*Azolla filiculoides*). It also produces 0.5 kg of methane.

A duck fattened on cane juice from brooding (700 g of liveweight at 21 days of age) to 2.5 kg of liveweight consumes about 18 l juice (20° Brix) and 3.2 kg of supplement.

Assuming that 1 200 ducks (four batches of 300) are to be fattened, they will consume 21 600 l of juice and the remaining 98 400 l from a total of 120 000 l will fatten 80 pigs.

The Sheep Subsystem

Two ha of sugar cane also produce 60 tonnes of tops. The cane tops, together with the gliricidia foliage, are sufficient to provide the basic diet for 29 African hair sheep (adult ewes), one ram and their progeny. In addition, the sheep consume 1 280 kg of molasses-urea blocks, 2 160 kg poultry litter and 222 kg rice polishings, inputs which must be purchased.

The annual lambing rate is 1.9 (the average litter size is 1.22 with 1.53 parturitions per year) and the weaning rate is 1.7. With an average growth rate of 100 g/day up to weaning and 80 g/day from weaning to a slaughter liveweight of 25 kg, the lambs are sold at 255 days of age. Annual sales of liveweight are 972 kg. Annual methane production from the sheep is estimated to be 100 kg.

The Biodigestor and Pond Subsystem

The cost of materials (tubular polythene sheet and accessories) needed to construct a biodigestor that will supply a family of six in cooking fuel is about US$100. The ponds that receive the effluent are also used to grow water ferns (*Azolla filiculoides*) which can provide up to 50 percent of the protein needs for the final growth stage of pigs, i.e. from 50 to 90 kg of liveweight. At any one time, a farm has about 30 pigs, requiring 240 kg of azolla daily. This quantity can be produced from a pond surface area of 1 500 m² (Becerra, 1991).

The Earthworm Subsystem

This subsystem is still in the development stage (Cruz, Preston and Speedy, 1989); nevertheless, preliminary results are encouraging. From 1 m³ of cattle manure, over one year the production of California red worms was 6 kg (211 g of protein) (Rodriguez and Salazar, 1991).

Fresh worms have proved to be an excellent complement to the azolla. The combination of the two feeds (50:50 protein basis) has been used successfully to replace 50 percent of the soybean meal in a cane juice diet for fattening chickens (Rodríguez and Salazar, 1991).

Ducks consume azolla even more readily than chickens and it can therefore be expected that similar rates of substitution of the protein supplement can be achieved. Research to demonstrate this is currently in progress.

Estimated Inputs and Outputs of an Integrated Mixed Farming System

Estimations de la consommation et des productions d'un système agricole mixte integre. Estimacion de los egresos e ingresos de un sistema agrícola integrado mixto.

Subsystem	Inputs	Outputs[1]
Biomass	2 ha sugar cane	60 tonnes (816 GJ) bagasse (fuel)
	0.14 ha *Gliricidia sepium*	
Pig fattening	80 weaner pigs at 25 kg	7 200 kg pig liveweight
	4 240 kg supplement	40 kg methane
Duck fattening	1 200 ducklings at 0.7 kg	3 000 kg duck liveweight
	3 840 kg supplement	12 kg methane
Sheep rearing and fattening	29 ewes, 1 ram	972 kg lamb liveweight
	1 280 kg MUB[2]	100 kg methane
	2 160 kg poultry litter	
	222 kg rice polishings	
Net liveweight/ha		4 160 kg
Methane: meat (liveweight) ratio		0.017

Saleable liveweight, fuel and methane.

MUB = molasses-urea blocks.

Productivity

The overall level of livestock productivity using sugar cane is high. In addition, there is a considerable quantity of biomass (consisting of bagasse) which is a potential source of farm-based energy (Preston and Echavarria, 1991). Even if the sugar cane yield were no higher than the world average (about 55 tonnes/ha/year), the livestock output would still be more than 2 000 kg of liveweight/ha/year.

Sustainability

The system as a whole is environmentally friendly and sustainable, building on the concepts of ecodevelopment and self-reliance. Almost all needs are farm-grown with a minimum of fossil fuel-derived inputs, and a surplus of biomass energy is provided.

Based on estimates taken from Crutzen, Aselman and Seiler (1986), it is calculated that the pig/duck and sheep units will produce, respectively, 52 and 100 kg of methane per year. This results in a methane:meat ratio of 0.017, compared with an average of 0.75 for pastoral systems.

Agrochemicals are not used: biodigestor effluent, manure from the sheep and humus from the earthworms supply all the required plant nutrients.

Dead leaves from the cane and trees form a continuous mulch over the soil surface, thereby improving fertility, fixing atmospheric nitrogen (Patriquin, 1982), probably oxidising atmospheric methane (Keller, Mitre and Stallard, 1990; Mosier *et al.*, 1991) and certainly preventing erosion.

Target Groups and Self-Reliance

The system is directed at, and has the greatest impact on, resource-poor farmers whose family members may all be provided employment. It is not a package, but rather a series of subsystems which can be introduced independently. The innovative feature of the system is that it is integrated in such a way as to maximise utilisation of available natural resources and minimise inputs. The technologies themselves are not innovative. All are known and have been applied commercially in other contexts. Many farmers in Colombia are introducing or using either some or all of the elements that make up this integrated farming system. FAO-assisted projects to transfer the technology are-under way in the Philippines (TCP/PIII/8954) and Viet Nam (TCP/VIE/8954) while others are being planned for El Salvador, Barbados and Trinidad and Tobago. Elements of the technology are already being applied on a large scale in Cuba (Figueroa, personal communication) and are in the development phase in Mexico (Alvarez, personal communication) and Honduras (Esnaola, personal communication).

The "self-reliant" feature of the technology is that it has given a comparative advantage to small, as opposed to large-scale, producers by virtue of using farm-produced resources derived from a rational and sustainable exploitation of the natural riches of the tropical environment - solar energy, soil, water, biological diversity and people.

Research Strategies for Development of Animal Agriculture

The Case for Research

Most scientists believe that research will support the development of animal agriculture. Unfortunately, this belief is not always shared by the decision-makers who allocate the resources. It is therefore of importance to demonstrate that research is justified by more effective development. An important element of a successful research strategy is to ensure favourable impacts social, economic and environmental (ILCA, 1992).

Demonstrating the economic importance of animal agriculture, particularly in developed regions, is not difficult. Livestock products, such as meat, milk, eggs and hides, account for more than one-half of the value of total agricultural production, including tobacco and other non-food crops (USDA, 1990). In most developing regions, the proportional value of livestock products is lower but still appreciable. As a proportion of total agricultural production, livestock products amount to about 22 percent for Southeast Asia, 25 percent for sub-Saharan Africa (not including the Republic of South Africa), 26 percent for China, 31 percent for West Asia and North Africa and 38 percent for South America.

These values do not include animal traction and manure for fertilizer and fuel, which partially substitute for the fossil fuel-powered tractors and chemical fertilizers used in developed regions. For sub-Saharan Africa, the value of animal traction and manure as fertilizer was estimated to be about one-half of the combined value of meat, milk and eggs (ILCA, 1987). Using this proportional value for traction and manure, the combined contributions from animal agriculture are approximately 35 percent of total agriculture production in sub-Saharan Africa (compared with 25 percent for meat, milk, eggs and hides alone). For other developing regions - especially Asia - the values of traction and manure are likely to be equally important.

Research supports the development of animal agriculture in many ways, perhaps the most important being the enhancement of livestock productivity which leads to a more efficient utilisation of available resources. This potential for improving productivity is dramatically illustrated by the differences in meat and milk productivity of cattle in developed and developing regions.

One-third of the world's cattle, which are found in developed regions, currently produce 70 and 80 percent of the global beef and milk supplies, respectively. These impressive differences in meat and milk production primarily result from the greater productivity of cattle per head in developed regions. The average yield per cow milked in developed regions is 3 605 kg compared with only 820 kg in developing regions (FAO, 1990). Similarly, the annual beef offtake per head of inventory in developed regions is about double that in developing countries - 35 kg compared with 18 kg.

Improved productivity benefits economic development both at the household and national levels. Livestock products improve the nutritional status of both farm and urban families. Sales of live animals, meat, milk, eggs and fibre are often the major income source for farmers of developing countries. Sales of milk and eggs provide a continuing flow of cash, a particularly important factor as farm families move from subsistence to cash-based economies.

Research can provide dramatic demonstrations of what is achievable and may thereby encourage development. Farmers, who are understandably averse to risks, are reluctant to make a major investment in their livestock when there is a high probability of losses through deaths. However, research that leads to effective vaccines and genetically resistant breeds can encourage farmers to invest in improved nutrition, management and other livestock interventions.

In recent years, global attention regarding environmental concerns has intensified. Too often and mistakenly, the development of animal agriculture is seen as being harmful to the environment. Research-based improvements in livestock productivity, on the contrary, should actually result in a more efficient and sustainable use of natural resources.

The central problem is that demand for food and fibre by the rapidly expanding human populations places great pressure on the often fragile natural resource base. This demand will increase, especially with regard to the food requirements of urban-based populations. The population of sub-Saharan Africa, for example, is expected to increase from the present 500 million to nearly 1 300 million by the year 2025.

Currently, only 145 million people are urban dwellers but, by 2025, more than 700 million people will be living in sub-Saharan cities and towns (World Bank, 1990). Urbanisation will cause a sharp increase in demand for nutrient-rich, easy-to-prepare livestock products. Meeting this demand will not be easy but, fortunately, research-based interventions should help.

Such interventions include increasing feed and residue yields from crops grown on arable lands, reducing losses from disease and parasites and improving the genetic potential for milk and meat yields. These improvements should have positive ramifications for long-term sustainability - both environmental and economic.

Research Strategies

Strategies for research were given particular attention in the recent study, *Assessment of animal agriculture in sub-Saharan Africa* (Winrock International, 1992). This study recognised that research is but one essential element along with extension, education and investment for the successful development of animal agriculture. In many cases, however, research can also act as a catalyst to expedite and ensure the success of development activities. Although the Winrock study focused on sub-Saharan Africa, the general principles and priorities mentioned are broadly applicable to animal agricultural research throughout developing regions.

A successful strategy must be based on the assumption that animal agriculture is a positive factor in long-term, sustainable agriculture and that research will enhance the contribution of livestock to sustainable agricultural development. Strategies should thus embody the following concepts: adequate economic returns to livestock farmers; maintenance of natural resources and productivity; minimal adverse environmental effects; optimal production with minimal external inputs; and satisfaction of human food and income needs as well as of rural families' social needs.

Farm Systems Approach: Given the complexity of animal agricultural management, a farm systems approach to developing and implementing research strategies is essential. To ensure that the research undertaken addresses farmers' perceived needs, such an approach must: characterise the farming system - identifying inputs, outputs, potentials, constraints and interactions of components; develop interventions to resolve constraints and exploit potentials; design alternative systems; evaluate interventions and alternative systems; and elaborate on alternatives that have proven to be technically practical, economically feasible and socially acceptable.

The research process should begin with the careful identification of problems. Systems description and constraint analysis are particularly valuable for informing scientists about the real problems faced by farmers.

An interdisciplinary team effort, combining efforts of socioeconomic and biological scientists, is required to ensure that the full benefits are realised from the research to the development stage. In the planning and priority-setting stage, the social scientists can provide ex ante projections of costs and benefits while helping to address equity issues that may arise, such as gender and reconciling the interests of the poorer farmers with those of the more prosperous producers. As research progresses, the monitoring of socioeconomic factors improves the probability of favourable social and economic impacts.

Since the objective of farm systems research is to improve traditional systems and, consequently, production, it follows that there wild be circumstances calling for major changes. For example, the six-fold increase in pig and poultry meat production envisioned for sub-Saharan Africa over the next 35 years (Winrock International, 1992), will require the adaptation of high-input, industrialised production systems to the environmental conditions of sub-Saharan Africa. This adaptive research will have little relevance to the problems and constraints of traditional backyard, scavenger poultry production.

Research Priorities

The research strategy recommended for sub-Saharan Africa identified the following priorities: feed supply, animal health, genetic improvement, livestock management, crop/livestock farming systems, natural resources and policy (Winrock International, 1992). Priorities for research were based on agro-ecological, biophysical and socioeconomic constraints and potentials. Criteria for establishing research priorities included: potential for achieving substantial gains in production and/or income; probability research to resolve constraints and/or exploit potentials in order that substantial benefits be realised; availability of readily applicable technology to resolve constraints and exploit development potentials; and the expected social and environmental impacts of research.

Feed Supply

The primary constraint to livestock production in sub-Saharan Africa is the fluctuating quantity and quality of the year-round feed supply. Ruminants will continue to depend primarily on forages and crop residues. However, energy and protein concentrates are required for the expansion of poultry and pig production while concentrate feeds will also be needed to supplement the diets of high-yielding dairy cattle. In the drier regions, seasonal shortages of forage are common while, in wetter regions, the nutritive value of forages varies seasonally and the failure to preserve surpluses therefore limits year-round carrying capacities.

Arid Zones: Low, erratic rainfalls preclude any significant increase in feed production without irrigation. Thus, research priorities concentrate on the more effective utilisation and maintenance of natural vegetation for which the requirements include monitoring systems to facilitate early warning and drought relief measures as well as grazing management schemes with an emphasis on social and organisational prerequisites for long-term sustainability.

Semi-Arid and Subhumid Zones: These zones offer the greatest opportunities for research-based improvements in feed supply. However, care must be taken to match interventions to specific ecological characteristics in order to ensure long-term environmental sustainability. Fortunately, feed production options such as forage legumes and multiple-purpose trees (MPTs) are often environmentally beneficial. Their use by livestock brings economic benefits to farmers as well as improving soil fertility and controlling erosion. Rain-fed cropping systems will expand in these zones in response to population growth and food demand and the better grazing lands that are not already cultivated soon will be. Therefore, research should be directed towards forage production on marginal lands; forage legumes and MPTs as rotational and complementary crops; improved utilisation of crop residues; strategic protein, energy and mineral supplementation to correct nutrient deficiencies and promote efficiency; and conservation of seasonal surpluses to compensate seasonal feed shortages.

Highlands: Since human and livestock populations are nearing the maximum carrying capacity of most African highlands, research should concentrate on more intensive systems that produce higher yields of protein and energy. Most crop/livestock farmers are small-scale with limited access to grazing lands, a situation that calls for integrated cropping systems combining food, cash and forage crops (including MPTs) to meet the needs both of livestock and humans while sustaining the soil and water resource base. Coarse grains, roots and agricultural by-products will be increasingly used for semi-intensive dairying; the feeding of cattle, sheep and goats for slaughter; and intensive pig and poultry systems. Ways to enhance and preserve feed values of crops must be sought, an endeavour that could involve collaboration between plant breeders, agronomists and animal scientists.

Humid Zones: Human population expansion and the subsequent growth in food and feed demand will encourage agricultural development in humid zones. There is concern about the implications this might have for the often fragile natural resource base (discussed later) and research efforts to increase and improve feed supply must have full cognizance of such environmental issues. Alley farming with MPTs is an example of a research-based alternative which can enhance crop productivity, protect the resource base and provide high-quality forage for livestock.

Poultry and Pigs: Industrialised commercial production of poultry and pig-meat will provide nearly one-half of the meat requirements for sub-Saharan Africa, estimated to be 8 million of the total 19 million tonnes of meat that will be required annually by the year 2025 (Winrock International, 1992). The demand for feed grains is predicted to increase from 3 million tonnes in 1990 to 25 million tonnes by the year 2025, and for protein meals from 0.2 to 6 million tonnes. If past experiences are repeated here, these concentrate requirements will be met from coarse grains, root crops and oilseed meals produced as cash crops over and above human requirements. Research must aim to fulfil a similar demand for food crops, but with a predominant emphasis on increasing yields under prevailing environmental conditions; rainfall, plant diseases and soil fertility, for example.

Animal Health

Diseases and parasites seriously limit live stock productivity throughout sub-Saharan Africa. Research is needed to improve the efficacy of existing preventive and therapeutic treatments and to develop new diagnostics and vaccines.

For some health problems, resistant genotypes and preventive management offer cost-effective alternatives to treatment.

Trypanosomiasis, tick-borne diseases (theileriosis, cowdriosis, anaplasmosis and babesiosis) and the tick-associated disease, dermatophilosis, were identified as major constraints to livestock production in sub-Saharan Africa (Winrock International, 1992). Priority was given to genertic engineering research into the development of highly potent and thermostable multivalent vaccines. Diagnostics, based on monoclonal antibodies and recombinant DNA technologies, must aim to identify infectious agents and ensure efficient epidemiological monitoring, while research on immune responses will provide the basis for developing effective vaccines and delivery systems. Similar research on diagnostics, vaccines and delivery systems is needed to improve the control of major epidemic diseases such as rinderpest, contagious bovine pleuropneumonia (CBPP) and pestedes petite ruminants (PPR).

Economic losses from diseases aggravated by intensified farming systems, including losses from abortions, pre-weaning mortality, internal parasites and mastitis, will become increasingly important. Integrated disease control strategies will therefore need to be devised while research should concentrate on diagnostic techniques, preventive management and genetically mediated resistance. The traditional emphasis of veterinary research on major epidemic and tick-borne diseases is gradually changing as the cumulative economic losses from less dramatic animal health problems become apparent. A greater understanding of the epidemiology and economics of animal health would assist in the appropriate allocation of resources and the delivery of health services (Winrock International, 1992.)

Genetic Improvement

Global concern about loss of biodiversity applies to domestic as well as wild populations. Genetic resources must be characterised and preserved and their diversity used to improve livestock productivity. Under the harsh production conditions of many developing regions, genetic adaptations to disease and climatic stresses are particularly important.

The need for research ranges from genetic manipulation at the molecular level to the crossing of high-yielding "exotic" breeds with well-adapted indigenous genetic resources. Fortunately, basic research at the molecular level, which is now under way in developed countries, can establish a basis for future applied research in developing regions. In addition, the principles of quantitative genetic theory may be transferred to genetic improvement programmes involving selection, cross-breeding and multiplication for example, multiple ovulation embryo transfer (MOET) and open nucleus breeding schemes. Thus, the identification, characterisation and development of indigenous genetic material should take priority in the allocation of the scarce resources available for genetic research. In Africa, these include the unusual, perhaps unique, genotypes of *Bos taurus* cattle in West Africa, African hair sheep, camels and donkeys. Major research initiatives with indigenous populations are being developed by ILCA and FAO. These will be largely implemented under contracts with scientists of national agricultural research systems (NARS) through the collaborative research networks discussed later.

Resource Management

The need for expanded agricultural production to provide food for today's population is increasingly affected by the fear that the eroding natural resource base will not meet the food needs of future generations. Farmers are already cultivating the better grazing lands, thereby limiting pastoral herds to marginal lands - even these marginal lands are increasingly threatened by the expansion of cultivation. If agricultural development is to be sustainable, research is needed to intensify and increase productivity from the better endowed and more robust lands and to improve the management of soil, water and vegetative resources on the more fragile lands (Winrock International, 1992).

Although environmental consciousness is high - regarding global warming, desertification, deforestation, for example - knowledge of causal relationships and their management is poor, especially in the case of animal agriculture. Long-term monitoring studies incorporating geographic information systems, remote sensing and modelling have become research priorities for most developing country environments supporting animal agriculture, especially as the inadequacies of paradigms derived from experience in temperate environments are increasingly evident (Winrock International, 1992).

Development of livestock production in the humid forest regions of sub-Saharan Africa is not recommended by the Winrock study. The adverse effects of tropical deforestation, such as climate change, loss of biodiversity and soil degradation have been well publicised. However, livestock production systems based on humid savannahs can be quite productive, providing control of major diseases, especially trypanosomiasis, is effected. Increased population pressure will almost certainly expand exploitation of the humid zone. The fallow period of traditional slash-and-burn farming systems is already being shortened in many regions, leading to the pernicious loss of soil resources. Research on nutrient cycling involving the use of animal manures, usually in combination with chemical fertilizer, may offer alternatives to slash-and-burn practices in the humid zone.

Policy Research

The policy environment can have either positive or negative effects on investment and innovation in animal agriculture. The absence of sound economic policies in support of animal agriculture will impede:

- investment in infrastructure;
- proper incentives to farmers;
- adequate supplies of production inputs and the delivery of animal health services;
- effective marketing and credit facilities;
- increased animal productivity through biological research.

Milk is a major source of income for smallholders. Traditional management practices gradually evolve as the growing demand for milk justifies investments in improved feeding, health, breeding and processing hygiene - Le fait est une source majeure de revenue pour les petite agriculture. Les pratiques de gestion traditionnelles evoluent progressivement a measure que la demande de fait justifie des investissements pour ameliorer l'alimentation, la sante et la selection des animaux ainsique l'hygiène du traitement du lait. - La leche es una fuente importante de ingresos pare los pequeños agricultores. Las practicas tradicionales de ordenacion se modifican gradualmente dado que la demanda de leche justifica la inversion pare mejorar la alimentacion, la sanidad, la seleccion genetica y la higiene durante el proceso de elaboracion

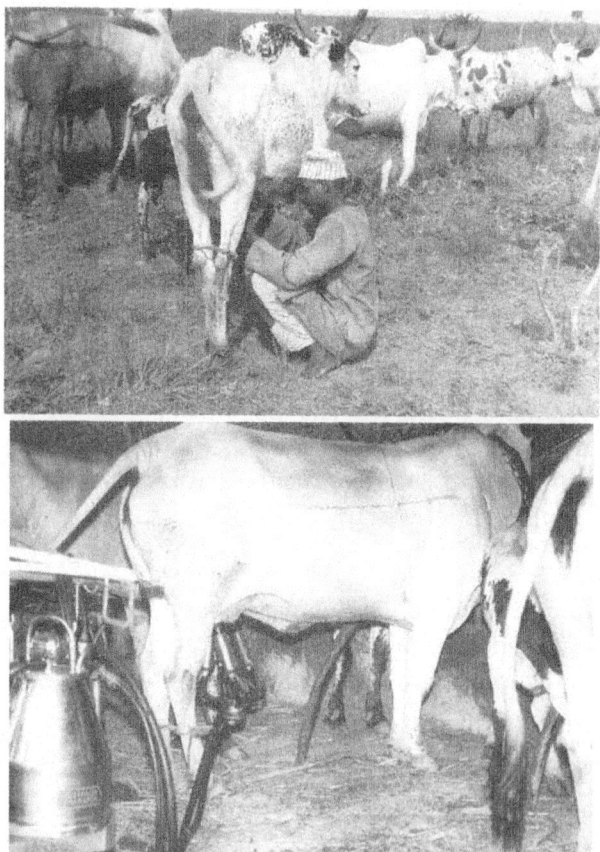

Figure 4: Cattle and other trypanotolerant breeds are important genetic resources for meat, milk and power in tsetse-infested regions of Africa

Policy research could improve data bases and provide analyses to help policy-makers anticipate and understand the probable consequences of their actions. The Winrock study identified policy research needs in the general areas of fragile land use, institutional policies and fiscal incentive and trade policies.

During a recent planning workshop, Livestock and Resource Management Policy, convened by ILCA, an international group of policy analysts and researchers identified the following research priorities: resource management; trade and macroeconomics; technology policy, markets and institutions.

Resource Management: A major constraint to the adoption of improved innovations in animal agriculture is the nature of various communities' claims to the natural resource base. Insecure tenure, multiple ownership, common property and a lack of clearly defined and secure property rights result in the overexploitation, underinvestment in and general mismanagement of resources. This is fuelled by a poor understanding of the appropriate role for institutions that govern the use of land, water, rangelands and other resources. There is therefore a need to study how resource management policies, including land-use rights, affect resource use and how changes in policies might advance environmental objectives.

For the mixed production systems of the semi-arid and subhumid zones, these studies might examine resource competition and complementarily between different land uses and enterprises. For the humid areas, analysis of the relationships between disease control, livestock development, resource use and the environment are required. For the pastoral production systems of the arid zone, the linkages between pastoral societies, rangeland tenure and rangeland ecology should be studied.

Farmer support for resource conservation has been shown to depend greatly on government policies. The effects of credit, pricing, and monetary policies on resource use and the environment, especially for the mixed production systems, are therefore important points to consider. Policy-induced distortions in financial markets, for example, result in small farmers having to pay high real interest rates. Many farmers who are unable to borrow or meet the repayments on borrowed funds seek refuge in common access areas, which are often susceptible to environmental degradation. Policy-induced distortions in agricultural commodity markets also result in farmers receiving low prices for their livestock, thereby discouraging them from investing in natural resource management. There is a similar linkage between resource management and exchange rate policy. Setting official exchange rates above market exchange rates discourages agricultural exports, including livestock. This in turn decreases the derived demand for land, which discourages farmers from managing existing farmland and developing new land in a responsible manner.

Trade and Macroeconomics: The macroeconomic environment plays a crucial role in the development of the national livestock sector. Macroeconomic policy affects opportunities for trade and may, therefore, affect expansion of the livestock sector. Among areas which merit attention for research are: the effects of structural adjustment or liberalisation on livestock production, concentrating on supply and demand changes resulting from economic incentives and constraints; the effects of and impediments to regional trade via economic integration; and the structure of demand for animal products.

Structural adjustment is already a reality for many sub-Saharan African countries, but while general impacts on the public sector may have been documented, specific effects on the livestock sector have not. There is a renewed interest in regional trade agreements and economic communities (such as the Economic Community of West African States and the Preferential Trade Area covering East and southern Africa).

Beneficial integration will require informed policy decisions, hence policy research is needed. Priorities include projecting changes in national comparative advantages for livestock production enterprises over time and identifying the "winners and losers" when existing trade impediments are removed. Although considerable work has been undertaken on food demand in sub-Saharan Africa, relatively little attention has been devoted specifically to demand for animal products. Understanding the effects of macroeconomic policy adjustments on the livestock sector requires a knowledge of demand structures.

Technology Policy, Markets and Institutions: Research topics identified in this area include how price and non-price factors influence technological change. For the factor markets, issues of interest include the effects of land tenure, credit and labour policies. For the output markets, issues include the effects of input and product prices together with non-price factors such as quality and infrastructure. Policies to strengthen NARS and extension systems will benefit from research on institutional change, government expenditures and their allocation, institutional structures and linkages and the efficiency of public services in the livestock sector.

Research Institutions

A comprehensive research strategy should utilise basic, strategic, applied and adaptive methods as doomed appropriate. The comparative advantage for undertaking these types of research generally varies depending on the institution. For example, NARS have the ready access necessary to address the specifics of local farm systems. Whereas, International Agricultural Research Centres (IARCs) are generally better placed to address strategic and applied research needs at the international level.

NARS: National agricultural research systems are the foundation of a successful national research strategy. To be effective, NARS should include linkages to extension services, academia and private industry. In aggregate, NARS constitute the majority of human, physical and financial resources that can be brought to bear on research problems.

For example, in sub-Saharan Africa annual funding for NARS (expressed in 1980 dollars) was approximately US$372 million during the period 1981-85 (Pardey, Roseboom and Anderson, 1991). This was about four times the amount allocated to IARCs in sub-Saharan Africa. Unfortunately, the resources available to NARS are rarely utilised effectively. Too often, the majority of resources are devoted to the maintenance of staff and infrastructure. Thus, external donor funding is required for the marginal costs of experimentation.

The NARS addressing animal agriculture in sub-Saharan Africa are organised in several modes: as separate semi-autonomous research institutes, as departments within ministries and university-based institutes or as independent departments. There is often little connection between plant and animal-oriented research, even in countries where mixed crop/livestock systems are predominant.

Stakeholders in the research process - farmers, extension agents, agribusiness, policy-makers - are not usually involved in setting national research priorities and developing programmes, but their involvement is essential for ensuring that research will be relevant to national needs (Winrock International, 1992).

Regional Organisations: Regional collaboration among NARS concerned with similar problems offers an important opportunity to accomplish more with the scarce resources available for research. In Africa, there are several regional organisations addressing animal agriculture. The Centre International de Recherches et Development sur l'Elevage du Zone Sub-humide (CIRDES), formerly Centre de Recherches sur les Trypanosomoses Animales (CRTA), located in Burkina Faso, has recently expanded its mandate for applied and adaptive technology transfer and training in livestock production and health. It has signed agreements with Benin, Burkina Faso, Côte d'Ivoire, the Niger and Togo. The International Trypanotolerance Centre (ITC), based in the Gambia, also plans to serve a broader regional mandate.

Networking allows collaborating NARS partners to pool scientific efforts on a regional basis in order to address problems of mutual interest more effectively, thereby avoiding a duplication of efforts. Successful collaborative research networks typically have the following characteristics (ILCA, 1991):

- a well-defined common theme and strategy;
- an existing or potential source of improved technology;
- a harmonising (coordinating) institution serving as the hub of the network;
- regular meetings of participating scientists;
- an information exchange system;
- free exchange of results and methods among members;
- education and training opportunities;
- financial support for in-country research activities conducted by national scientists;
- explicit national commitments for research on the commodities covered by the network.

Multilocational regional projects, managed through networks, offer considerable opportunities for enhancing the efficiency of research. They allow the introduction of standardised methodologies and hence lead to more significant conclusions than can be obtained from isolated experiments.

IARCs, such as ILCA, can play a major role as collaborative research partners in networks, providing training opportunities for network participants, disseminating research methods and results and facilitating the exchange of information.

IARCs also assist with network support functions which include helping to attract donor funding, helping to organise networks (setting up network steering committees), sponsoring meetings of participating scientists and providing services in areas such as data analysis, documentation and publishing.

IARCs: Research activities at the international level include: assessing the changing research needs of global agriculture, fisheries and forestry; the collation, processing and dissemination of scientific information; the collection, preservation and exchange of germplasm and improvement of methodologies for its use; the enhancement of germplasm for crops, livestock, trees and fish dominant in the economic activity of many countries; the development of resource management and husbandry principles appropriate for agro-ecological conditions widely distributed around the globe; strategic research on production processes; and specialised training (TAC/CGIAR, 1991).

IARCs, supported by the Consultative Group on International Agricultural Research (CGIAR), are expected to conduct strategic and applied research of an international character that complements and supports the efforts of NARS, their principal clients. The ultimate objective of IARC is to benefit the poor in developing countries through technological change leading to increased food production and income generation. Among the 16 IARCs currently supported by CGIAR, two are primarily devoted to animal agriculture - ILCA and the International Laboratory for Research on Animal Diseases (ILRAD).

The International Centre for Tropical Agriculture (CIAT) in Cali, Colombia, and the International Centre for Agricultural Research in the Dry Areas (ICARDA), Aleppo, Syrian Arab Republic are primarily oriented towards crop research but devote significant resources to livestock-related research: CIAT to tropical pastures ICARDA to small ruminants. The International Food Policy Research Institute (IFPRI), Washington, D.C., has identified animal agriculture as a promising basis for economic development as well as for policies which support animal agriculture.

The International Service for National Agricultural Research (ISNAR), the Hague, the Netherlands, has encouraged the strengthening of NARS capacity to generate appropriate animal production and health technologies. In the past, ISNAR has emphasized capacity-building for crop-oriented research. However, more attention should be given to strengthening NARS to address crop-livestock and animal-based systems.

An advantage of IARCs research is that results "spill over" into similar agroecological and socioeconomic conditions across national boundaries. IARCs are also well positioned to assist NARS with the transfer of results from basic and strategic research by specialised research institutes in developed countries.

Technology Development

Technology is a major research output for the development of animal agriculture. It has been argued that an ample stock of appropriate technology is already on the shelf, but experience has generally shown this not to be true.

Technology Transfer: Only rarely are technologies from developed regions directly transferable to developing regions. Climate, availability of services, trained personnel, dependability of power supply, marketing infrastructures are but a few of the factors affecting how well technologies work. These factors usually differ markedly between developed and developing regions. Thus, strategic, applied, adaptive and even basic research initiatives are required to adapt technologies to fit needs of animal agriculture in developing regions.

Research to adapt and transfer technologies for small-scale farming systems will primarily involve publicly supported institutions, NARS and IARCs. Privately-funded initiatives are more likely to address the needs of the industrialised swine and poultry systems as well as the development of those technologies which are readily marketable.

Appropriate Technology: The concept of "appropriate technology" is often interpreted to mean that technologies suitable for developing countries are less sophisticated and advanced than those used in developed countries. It follows, therefore, that a research strategy to develop appropriate technologies for developing countries will more than likely not involve advanced science. This generalisation frequently holds and the highest priority should therefore be given to adapting relatively "low-tech" interventions. There are, however, important exceptions. For example, the battery-powered notebook computer - certainly in the forefront of available personal computing technology - is particularly useful where there is an erratic power supply, as is the case in many developing countries. Research strategies embodying development and application of advanced technologies should not be dismissed out of hand. Specific examples where state-of-science research is particularly appropriate to animal agriculture in developing countries include:

- molecular genetics research, including genome mapping and genetic engineering to combine productive and adaptive traits (Brem and Wagner, 1991);
- development of thermostable, multivalent vaccines and animal-side diagnostics for field use;
- reproductive technologies, including in vitro fertilization, embryo transfer and other techniques to assist characterisation of indigenous genetic resources;
- fermentation technologies to facilitate feed and food processing and preservation; for example, biocontrol of fungi and other microbes which cause silage spoilage and loss of nutrients;
- development of transgenic rumen bacteria to enhance cellulolytic activity and detoxification of antinutritional factors in foodstuffs.

Strategies for Sustainable Animal Agriculture in Developing Countries

Animal agriculture is a complex, multi-component, interactive process that is dependant on land, human resources and capital investment. Throughout the developing world it is practised in many different forms, in different environments and with differing degrees of intensity and biological efficiency. As a result any meaningful discussion of the subject must draw on a broad spectrum of the biological and earth sciences as well as the social, economic and political dimensions that bear so heavily on the advancement of animal agriculture. There is a growing consensus among politicians, planners and scientists alike that livestock production in the third world is not developing as it should, or at a sufficient pace to meet the high quality protein needs of a rapidly expanding human population.

The sobering reality is, despite the many development projects implemented over the years by national, bilateral and multinational agencies and often substantial capital investment, there has been little or no change in the efficiency of animal production in the developing world. Livestock numbers have increased substantially in many countries and while the growth in output is welcome, it does not necessarily equate with sustainable productive growth. On the contrary it can, as it has done in the drought prone arid regions, lead to a lowering of productivity and degradation of the rangelands.

The purpose of the Expert Consultation was to discuss and formulate specific criteria and questions relating to the planning and implementation of sustainable livestock production programmes in the developing world. There is increasing concern regarding the conservation of the natural resource base and protection of the global environment and FAO attaches highest priority to the sustainable development of plant and animal agriculture.

This Expert Consultation is one of a number of initiatives being undertaken by FAO to ensure the sustainability of it's agricultural development programme.

The discussion and recommendations arising from this Expert Consultation have been used to help to focus and guide global, regional and national policies and action programmes on the sustainable development of agriculture and have provided an important contribution to the FAO/Government of the Netherlands International Conference on Agriculture and the Environment held in the hague, 15–19 April, 1991.

The Consultation consisted of three days of plenary sessions and two days of workshops and was structured so that it would address the following main areas of concern:
   • development policies and environmental issues,
   • practical technologies, and
   • support services to animal agriculture.

The purpose of this paper is to outline the technical issues and major policy considerations related to the sustainability of livestock development. The identification of these issues is primarily based on the manifestations of unsustainability in certain patterns of livestock production in developing countries. For the purpose of this paper sustainability of livestock development will be looked at in the context of management and conservation of the natural resource base, i.e. animal and feed resources. The environmental and socioeconomic considerations underlying sustainable development are elaborated in FAO (1989).

Livestock development over the past three decades has been mainly directed towards satisfying the rapidly increasing demand for milk, meat and eggs in the urban centres. The required intensification of animal production was often limited by both an inadequate indigenous feed and animal resource base along with a fragile socioeconomic capacity in most developing countries. In many cases, the political pressure exerted by the urban demand led to the importation of exotic stock and concentrate feeds. Technologies which proved successful in the socioeconomic context of developed countries were applied. Technical institutions were also founded on similar rationale. Large amounts of public funds were used to support the modern production systems. The traditional livestock sector was also stimulated by the increasing prices for animal products but its growth was hampered by inadequate input supply and support services.

The signs of unsustainability of such development began to surface in the past few years. The sudden increase in grazing pressure and lack of capacity for conservation among the graziers led to the degradation of rangelands.

Feed imports could not be sustained. Lower adaptability of imported stock did not allow rapid multiplication and crossbreeding raised the fears of erosion of the indigenous genetic diversity. The pollution of water and air has been limited to the few areas of livestock concentration in intensive production units. The large-scale use of insecticides in the past for Tsetse control and sporadic pollution of waterways with acaricides may have had significant environmental consequences.

Apart from the manifestations of unsustainability, the overall increase in animal productivity in the past three development decades has been minimal. Increased external assistance *per se* did not have a lasting impact on the performance of the livestock sector in developing countries (Bommer & Qureshi, 1988). Particularly, smallholder or rural livestock production did not pick up any significant momentum. Major issues underlying these generalised observations are outlined in the following sections.

Demand-Driven Development And Sustainability

Development Patterns: The high economic demand for milk and meat in developed countries has led to the establishment of capital intensive systems that require high animal productivity, high levels of concentrate (grain) feeding and highly mechanised or automated infrastructure. The production technologies developed over the years have successfully met these requirements. The problems currently faced by these systems arise basically from economic pressures and revolve around capital costs, input costs and subsidies. There are also environmental problems associated with high concentration of stock, animal wastes and high levels of fertilizer use for pastures.

Feed availability is not a constraint in feed deficit areas thanks to the well-established trade in feedstuffs. Large quantities of feed are imported from developing countries. In 1988, some 3.7 million tons of cereal brans and oilseed cakes were imported in the developed countries from the feed-deficit developing countries.

The prevailing production systems in the developing countries are a world apart. Large numbers of ruminant livestock graze common grassland especially in Africa. This low-input system had, in the past, successfully met a large part of the demand for meat in urban centres and catered for the local demand for milk and meat. For the past 2–3 decades the system has not been able to meet the increased demand arising from population explosion, increased income and rapid Urbanisation.

Livestock numbers on common grazing areas have increased way beyond the carrying capacity. The lack of economic capacity among the graziers to undertake range improvement resulted in degradation of most of these grasslands. After an increase in numbers and offtake during the past 2–3 decades, the present output is declining. The degraded common grasslands cannot be rehabilitated without massive investment and the politically difficult land tenure policy.

A sizeable proportion of the third world livestock is also kept in arable farming areas and is integrated in the system to provide draught power, fibre and dung as well as marketable quantities of milk, meat and eggs. A few holdings in these areas are large enough to warrant increasing the producing ability of the stock. On small farms, the large ruminants are kept primarily for work; milk and meat are by-products. Small ruminants are primarily kept for meat; milk is a by-product. Crop residues are the main feed resources although a small part of the land is devoted to fodder for work animals. Poultry and pigs basically scavenge the farm and household wastes. Supplementary feeding depends on the market for produce. In recent years, the increased price of milk and meat and improved marketing opportunities for the small farmers have, in some cases, resulted in increased fodder cultivation, better utilisation of crop residues and by-products, supplementary feeding and animal husbandry geared towards marketable milk, meat and egg production. However, except for poultry, these improvements have not taken root to a significant extent.

Spurred by high demand for milk, meat and eggs in urban centres, heavy investments were made in dairy farms, ranches and feedlots following the developed country model. This was facilitated during the mid-seventies to mid-eighties by growth in external loans, foreign exchange earnings and availability of capital at low interest rates. In most countries which encouraged this model the available foreign exchange was used to import high yielding stock and feedstuffs. However, the contribution of these enterprises to total milk and meat supply has been minimal and at high cost. With increased capital and foreign exchange shortages in recent years, the specialised livestock farming enterprises and feed industries are trying to re-orient their operations towards greater reliance on locally available feedstuffs and animal resources (FAO, 1990a). There is a growing realisation of the value of multipurpose local stock and their capacity to utilise fibrous feedstuffs.

A major part of milk and meat in developing countries is produced by the resource-poor small farmers. Smallholders contribute by far the largest part of the labour force. To improve production and income, techno-economic changes must be brought in the small farming system. Availability of sufficient feeds of adequate quality is the basic constraint in the system. Possibilities of providing external feed inputs are limited. Due to a lack of sufficient attention to the smallholder, appropriate technologies to improve performance of locally available animal and feed resources within the rural system have not developed. Institutional capacities in developing countries have also not been built up from this viewpoint. Smallholder dairy development, for example, has succeeded only when appropriate production technology was supported by an integrated dairy development programme incorporating milk collection, input supply and favourable price policy (FAO, 1984).

Technology Requirements

The objective of livestock development at the national level has been to attain as much self-sufficiency as possible to satisfy mainly the urban demand. At the farm level, the objective is to increase income and utilise family labour year-round. The role of animal production technology in developing countries revolves around finding labour-intensive procedures which maximise production with low-cost inputs. Feeds comprise by far the largest component of input cost.

The efficiency of feed utilisation in terms of herd or flock output is the primary consideration in technology development. Other elements of the development of appropriate and affordable technologies that ensure sustainability are: conservation and improvement of resource base; minimisation of wastes and environmental degradation; and recycling of wastes for animal feed or biogas.

Direct transfer or transplantation of technology from developed to developing countries has rarely been successful and sustainable in achieving the above objectives. However, in the case of a few specialised enterprises where such a transfer has been successful, the longer term consequences have been dependent upon imported feeds and exotic animals in order to take maximum advantage of the transferred systems. The side effects of imported technologies has been the neglect of indigenous livestock and feed resources (Preston and Sansoucy, 1987).

A sustained growth of indigenous production of milk and meat is dependent upon the introduction of new technologies that would be adopted by producers, especially, the smallholders. A systems approch is indispensable for the development of such technologies.

This approach should incorporate the rigour of the scientific method as well as the human element of involving the producer of the farmer at all stages of development. There is a pressing need to utilise the effective tools and techniques that are available to pursue a farming systems research and extension strategy. The farmer must be a key partner in every effort in the improvisation and introduction of the required technologies.

Multi-disciplinary effort and effective linking of research, extension and training are important requisites for the systems approach. The process of increasing animal productivity must also be effectively linked with other developmental elements such as market considerations, produce organisation, incentives and policy-making. Vigorous research-extension efforts have proved to be the basic requirements for improving animal production systems.

During the past three decades, most developing countries have been able to establish institutional infrastructure for livestock development, e.g. research centres, extension services, veterinary laboratories, disease control services and educational institutions at various levels.

The technical performance of this infrastructure is variable from country to country and from institution to institution. On the whole, the institutional impact on livestock production has been open to question. The development concern these days is not so much about the capacities in terms of physical infrastructures or size of trained manpower but about the usefulness of this capacity in improving farm output. The underlying constraints emerge from the prevailing traditional, neo-social, economic and political environment (Bommer & Qureshi, 1988).

Self-reliant livestock production in developing countries requires a strategy to optimise production from available feed resources through an integrated technology which employs multi-purpose crops, multi-purpose animals and recycling of residues and by-products. The identification of needs and a careful study of feed and animal resources are essential first steps. Resources must be examined in the context of various agro-climatic zones and fodder crops which might be grown. Ruminant production systems must then be matched with the resources in a way that aim for economic rather than biological maximisation. To introduce new technologies it is important to start with on-farm improvements. The utilisation of locally available feed resources must be maximised to reduce or eliminate the importation of concentrate feeds (FAO, 1985).

Livestock development efforts in the past laid primary emphasis on rapid genetic improvement arguing that improvements in feeding will be ineffective when animals with low genetic potential are raised. In recent years there is a growing consciousness to balance the rate of genetic improvement with improvements in feed availability and management. There is also an increased realisation of the potential of indigenous cattle, buffaloes, sheep and goats as multi-purpose animals suited to sustainable production systems and as efficient convertors of locally available feed resources. Innovated breeding and management procedures have not been effectively implemented to improve reproduction and to increase the milk, meat and work outputs of the indigenous stock.

Major Issues

Technology Development: Developing technologies for sustainable animal production has always been the objective of animal scientists even this was not expressly stated. This concern had been well taken into account regarding the environment in which the technologies were developed. However, the use, rather than misuse or misplacement, of a proven technology could be an issue in the context of sustainability in different environments. The important issue in this regard is the development of appropriate and affordable livestock technologies suited to specific agroclimatic zones in developing countries. To satisfy the criteria for sustainability, these technologies should support agricultural development which "conserves land, water, plant and animal genetic resources, and is environmentally non-degrading, economically viable and socially acceptable" (FAO, 1989). This would mean further research and development effort for the expansion of the feed base, conservation of animal geneti resources, recycling of wastes and efficient feeding systems.

Methodology to evolve sustainable production systems requires a multi-disciplinary approach and development of appropriate indicators of medium-term or long-term sustainability. These indicators should be practical enough to be used in project formulation, monitoring and evaluation.

Sustainable Institutional Support

The institutional framework for improving production systems needs to be structured, or restructured, to ensure multi-disciplinary collaboration and cost-effectiveness. Participation and cost-sharing by the farmers' organisations may be necessary in most cases. Present institutional structures for research, extension and veterinary services need evaluation.

Import/Export and Price Policies

Past experience has highlighted the unsustainability of livestock production and feed imports in most developing countries. The impact of imports on the existing production systems needs to be fully understood before allowing the import of milk, meat, feeds or live animals. Similarly, the export of agro-industrial by-products from feed-deficit countries disrupts livestock development efforts.

Price policies for milk, meat, feed or eggs should be geared towards creation of an economic environment in which it would be profitable to conserve and utilise local feed and animal resources for production. These policies should be devised and implemented for the benefit of the producer.

Degradation of Grazing Lands

Common grazing lands are most vulnerable to degradation. The basic issue in rangeland rehabilitation is land tenure policy which would encourage range improvement and proper grazing management. Access to the rangelands by the nomads and other poor segments of the population is also an issue.

Conservation of Indigenous Breeds

The relevance of the conservation and genetic improvement of indigenous animal genetic resources is well recognised (FAO, 1990b). The important issues concern the cost and the cost-effectiveness of methods to be employed. It has often been mentioned that these efforts should be supported by public funds and external assistance. However, this point of view should not divert the attention from efforts to develop privately-supported and cost-effective methods for conservation.

Methane Emission by Ruminants

There are various estimates of the contribution of ruminants to methane production and global warming. The reliability of these estimates apart, the issue concerns the options available to reduce methane emission. One option on which greater concentration of efforts may be needed is that of devising practical feeding procedures to reduce methane production per unit of milk or meat output.

The sustainable development of animal agriculture, across the many different ecoregions or agroecological zones that are found in the developing countries, poses many fundamental challenges; challenges to the primary users of livestock, to their extension, research and support service agents (private- and state-sponsored), to local and regional development authorities, to investment banks, government policy makers and their institutional organs and, in the final analysis, to the consumer or user of animal output. These challenges have been confronted by mankind with varying degrees of intensity since man first captured and domesticated wild animals for use. However, it is only in the past 200 years that what may be described as an industrial approach to animal agriculture became manifest. Over the past 40 years, the United Nations (UNDP and FAO), investment agencies such as the World Bank, and many bilateral and non-governmental organisations, all in consort with national governments, have launched development programmes to promote or effect the advancement of livestock production throughout the developing world. There have been many successes, particularly in pig and poultry production (e.g.in Thailand) but, unfortunately, many failures also.

Extensive reviews of livestock investment projects conducted by the World Bank (1985) and the Asian Development Bank (ADB, 1991) paint a gloomy picture of the success of development investment in livestock, particularly in the smallholder sector which farms the vast majority of all livestock in developing countries.

This lack of success/impact, in turn, has led to a significant reduction in the amount of investment funds directed to livestock production. Added to this, there is increasing interest in all issues related to the sustainability of agricultural development programmes, including livestock production, not only those factors that may directly effect the environment (e.g. Co_2, methane), but also a more thorough scrutiny of development inputs and technologies in terms of longterm practical application and impact.

The FAO definition of sustainability, embracing both dimensions of long-term development impact, provides the basis on which development strategies are discussed in this paper. Clearly there is a need to thoroughly analyse the determinants of sustainability in the context of animal agriculture so as to identify and evaluate effective development strategies.

Determinants of Sustainability

Effective development planning entails a comprehensive understanding of man's need to advance development, a thorough analysis of the technical, economic and social implications of proposed interventions and finally, an assessment of potential environmental impact (long-and short-term) that may influence local, regional or global conditions.

These very generalised determinants of sustainability, as represented, set a framework within which the development planning process must proceed. It is not the purpose of this paper to discuss livestock development planning at this broad level other than to highlight that complex interactions and conflicts of interest can, and very often do, exist among the different components in the development process.

The development audiences, i.e., the individual farmer, the extended family or local community and the regional and/or national governments, often have very different perceptions of what development means. Technical interventions may be viable economically, but may not be self-supporting (due to dependence on imports and availability of foreign exchange) and may not be socially acceptable if, for example, they result in increased or conflicting work routines within the farm family.

Finally, whereas few farmers in developing countries are acutely concious of the global environmental impact of their actions, they certainly are aware of their local habitats and, generally speaking, respect traditional practices and taboos that condition their farming patterns.

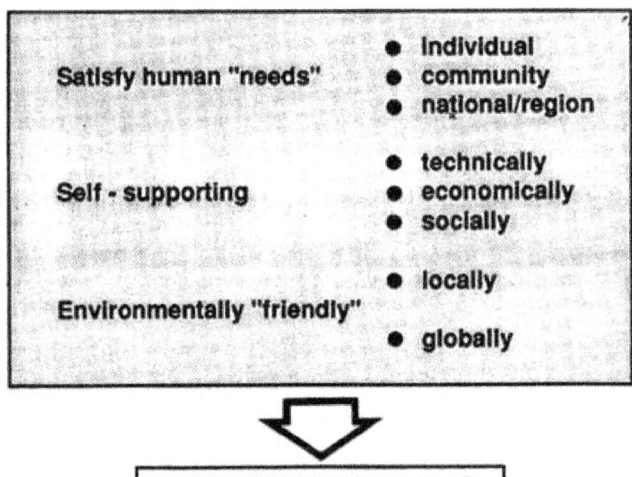

Figure 5: Determinants of Sustainability

Given the paramount importance of the broad determinants of sustainability as listed, and working on the assumption that these are duly considered in framing the development planning process, the main discussion in this paper focuses on technical strategies for the advancement of animal agriculture. These will be considered in reference to the classically-defined agricultural development resources, viz., animal and land resources, labour, capital and human enterprise. Issues will be discussed in a selective 'by way of example' manner, since a thorough discussion of any one issue is deserving of a full paper in itself.

The Animal: Production Objectives And Genetic Advance

The development of intensive livestock production in industrial countries over the past 30 years has become synonymous with single purpose breed specialisation; consequently, Holstein Friesians for milk production and continental beef breeds such as the Charolais, automatically spring to the minds of livestock development planners throughout industrialised Europe, America and Australasia. Unfortunately, some of these people often export their 'vertical thinking' into livestock development planning in many developing countries; countries to which specialised, single-purpose breeds are not suited and can rarely adapt to the harsh local environments.

Market demand, labour and capital costs, together with land availability, have dictated specialised single-purpose animal production systems in the industrial countries; production efficiency is ultimately measured in terms of shelf price and continuity of a standardised, high quality, animal food product.

However, interesting contrasts are found within industrial countries, if, for example, sheep management systems in a land-rich, low population country such as New Zealand, are contrasted with a high population European country such as Britain. In New Zealand, sheep breeding goals are focused on relatively low producing (lambing percentage, 90%-100%), easy-care sheep, where one man can manage approximately 4,000 ewes. In contrast, British lambing percentages in the smaller flocks (200-400 ewes) are targeted at levels approaching 200%.

However, the contrasts in animal breeding objectives between industrial and developing countries are much more stark. Most developing countries exploit the multipurpose use of livestock to produce meat, milk, fibre, draught energy and manure, as depicted. Several livestock development projects have failed to attract committed farmer participation simply because they were targeted on single-purpose milk or meat output and ignored the farmers primary interest in livestock as providers of animal traction or manure (Lensch, 1985). Further, the bioenergetic efficiency of multipurpose livestock production is often overlooked by livestock planners and, consequently, misplaced development objectives are often pursued.

In the Ganges delta of Western Bengal, for example, Odend'hal (1980) studied the energy inputs and output of 3,770 cattle on 2,600 smallholder farms which practised a typical Indian village livestock farming system. Total energy output from the livestock system (Kcal/year) as summarised in table, clearly shows that animal manure was a far more important output than milk or meat (calf weight).

This unique example is quoted, not to diminish the general importance of meat and milk in most developing countries, but simply to highlight the need to closely examine how and why farmers use their livestock before deciding on breeding or livestock development objectives in any given situation.

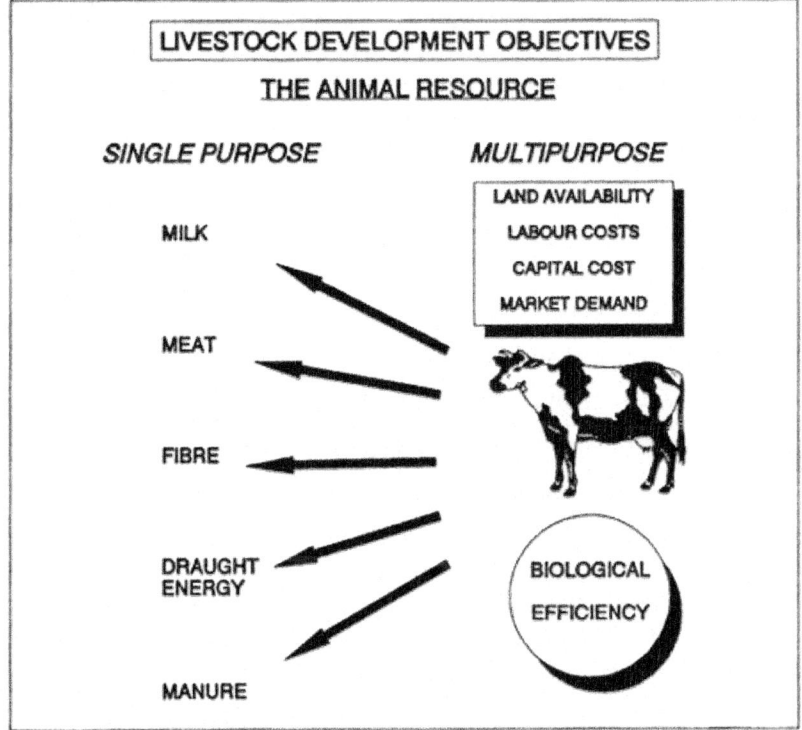

Figure 6: Livestock Development Objectives

Table 1: Total energy input and outputs in a traditional village cattle system in West Bengal

	Kcal/year
Manure	3.93 × 109
Traction	0.56 × 109
Milk	0.18 × 109
Calves	0.01 × 109
Total output	4.68 × 109
Total input	19.83 × 109
Efficiency	23.6%

Genetic Improvement

Perusal of the World Bank's analysis of livestock development projects over the past 25 years clearly shows that many projects focused their attention on the genetic improvement or, in some instances, the replacement of the indigenous animal resource. It is now widely accepted that the vast majority of such projects failed, and that projects aimed to promote exotic breed propagation failed miserably. Retrospective analysis of these projects would suggest that the fundamental tenets of effective genetic improvement were not adequately considered when these breed improvement projects were designed. The fundamental conditions that determine the success or failure of genetic improvement programmes are discussed here under four headings:

- Relevance of genetic improvement;
- Breeding objectives;
- Breeding strategies and plans;
- Genetic improvement capacity.

Relevance of Genetic Improvement

The concept of genetic improvement is attractive. However, reality suggests that improvement of the inherent genetic capacity of any animal population beyond the scope of the nutritional- or diseaselimited environment in which the population lives can be meaningless and is often counter-productive. Reports on several development and investment projects, sponsored by UNDP and the World Bank among others, offer tangible proof of this reality. Retrospective analyses of these projects raise the question of when genetic improvement is relevant and potentially useful. McDowell (1989) and Timon and Baber (1989) argue that it makes little sense to initiate genetic improvement programmes in livestock populations where average annual feed intake is less than 1.5 times the animals maintenance feed requirements; ideally it can be argued that feed intake levels should be twice maintenance. The feed intake levels in question refer to the population targeted for improvement, rather than just a nucleus population on a government breeding farm.

The relevance of genetic improvement programmes must also be judged in reference to the adequacy of the breed improvement infrastructure, not only to service and support genetic evaluation and selection within government-supported farms, but also in reference to the effectiveness of regional and national infrastructures through which the propagation and dissemination of the improved germplasm can reach the farmer.

This is often very weak or non-existent in developing countries; the failure to develop efficient artificial insemination (AI) services throughout most of the developing world is just one example of weak infrastructure. However, the most important criteria by which the relevance of breed improvement programmes is determined is farmer acceptability; do the changes brought about by a programme adequately benefit the farmers, in terms of perceived market benefit, investment risk, support service availability and general family welfare.

Breeding Objectives

Most animal breeders and, in particular, breeders of ruminant animals, identify breed improvement with increased levels of market-oriented productivity, be that meat or milk. This rationale has emerged from the more advanced countries where input costs (land, labour, capital or support service costs) in animal production are high. Unfortunately, these same breeding objectives, based on the premise that 'big is beautiful', have been exported to many developing country breeding programmes.

World Bank (1985) and ADB (1991) project reports provide many examples of failed attempts to achieve 'instant genetic improvement' by importing single-purpose high-producing breeds into unsuited environments. In the tropics, for example, selection objectives must identify animals that perform well under heat stress and can cope with the seasonally available feed supply and variation in feed quality.

Animals must not be evaluated or selected under better managed high-feeding conditions, which are sometimes developed on government farms or research stations, if these conditions are not typical of the prospective farming systems for which the animals are being bred. The classic experiment carried out by Falconer (1960), albeit with experimental animals, highlights this point.

His experiments showed that selection for growth rate on an *ad-libitum* plane of nutrition increased appetite, whereas selection for growth rate on a restricted diet increased efficiency of feed utilisation.

When genetic improvement within a harsh or limiting environment such as in the tropics of West Africa is considered, high output cannot be considered as the sole criterion of the animals' merit. Brody (1945), comparing a range of animal breeds, and indeed animal species, in terms of productivity per unit of metabolic weight, suggested that there were very small differences between big and small animals in net production efficiency.

Taylor et al (1985) largely confirmed Brody's thesis in a set of well-controlled experiments with different breeds of cattle. Comparable experimental evidence from small ruminants is not available but the conclusion is unlikely to be different. In other words, big is not necessarily beautiful in the context of net production efficiency.

It is now well established that specialised single-purpose breeding objectives should not be applied to livestock breeding in the difficult environments of developing countries. Indeed, it may well be questioned in the advanced countries at present as food mountains grow, as market buoyancy and prices are under pressure and as product to energy, feed and labour cost ratios continue to decline.

The basic question at issue is the biological efficiency of multipurpose breeding to produce meat, milk, fibre (skins) and, where relevant, draught power, as compared with the specialised production of one product.

Clearly, in the absence of factual data, it is not possible to generalise on this question but it is important to focus attention on the need for a much more comprehensive evaluation of production objectives when livestock breeding strategies are being determined, and particularly when considering difficult environments where economic and social values are very different from those in the more industrialised countries.

Breeding Strategies and Plans

The fundamental basis of a genetic improvement programme, properly focused, depends on selection intensity, selection accuracy and 'population reach'; population reach embraces the relevance of selection objectives and the efficacy of the propagation infrastructure within which the improved germplasm is disseminated to the target farmers.

Crossbreeding: As stated earlier, many genetic improvement programmes have, in the past, been based on the importation of exotic germplasm followed by the initiation of crossbreeding programmes in one form or another; the objective being to harness additive genetic variance in upgrading the indigenous population or the exploitation of heterosis in a continous crossbreeding programme. Choice of exotic breed is certainly important, but extensive reviews by Vaccaro (1990) in South America and McDowell (1983) in Asia raise serious questions about the adaptability of all exotic breeds introduced from temperate countries to tropical conditions.

Certainly, the proportion of 'exotic genes' introduced into indigenous populations is often debated, but the overall benefits of crossbreeding must be seriously assessed and measured, not just on the basis of one trait performance (e.g., lactation yield), but in terms of overall lifetime herd/flock productivity. Measured in these terms, there are few examples of successful and sustained crossbreeding/ upgrading programmes in developing countries.

The exploitation of heterosis in continuous crossbreeding as opposed to upgrading programmes, is often advocated as a way to rapidly advance animal genetic potential in developing countries. Certainly, there is good evidence of large heterotic effects on some components of animal productivity when exotic temperate and indigenous tropical breeds of cattle are crossbred (Cunningham and Syrstad, 1987). Despite this, and for very understandable reasons, sustained crossbreeding programmes are not evident anywhere in the developing world.

The explanation for this has as much or more to do with land use and economics than with genetics. Continuous crossbreeding systems involve three genotypic populations, viz., Breed A and Breed B interbred to produce the crossbred population A × B. In situations where natural mating prevails, this usually requires a sustained economically viable interaction of three sets of farming systems/breeders viz., one set each who breed populations A and B, and a third group who interbreed and farm the A × B crossbred. The sustained integration of three sets of breeders/farmers demands that their respective breeding practices fit to their land resources and compensate them accordingly.

Where efficient AI services are available, the basic issue remains the same, except that a sustainable crossbreeding programme with AI demands the complementarity of only two land-use systems instead of three. It is perhaps worth emphasizing that sustainable crossbreeding systems are rare in industrial countries and, indeed, only exist where they 'biologically fit' into traditional land use patterns, e.g., a lowland-upland-mountain land-use pattern as exists in Britain.

Molecular genetics (cloning coupled with embryo transfer) may offer new opportunities to exploit crossbreeding in the distant future, but in the immediate term (5 to 10 years) it can be argued that attention should be focused on the genetic improvement of indigenous breeds; in other words, selection within the indigenous breed population, if this is feasible.

Indigenous Breed Improvement: The basis on which to determine the genetic merit of an animal and to make accurate selection decisions have long since been established. Equally, the relative accuracies of different forms of selection, be they pedigree selection, performance testing, progeny or other forms of indexed family or combined selection, are well established. The absence of well-documented genetic parameter estimates makes it difficult in most situations to develop reliable selection indices for livestock breeding in developing countries.

However, this is not the only point to be considered. In the majority of countries, the infrastructure and support services necessary to collect the records from which accurate genetic rankings can be made, simply do not exist. Allied to this is the even greater problem of the very high level of phenotypic variation usually observed in these populations (Sands and McDowell, 1978, McDowell, 1983).

Coefficients of variation for phenotype can range from 30–60%. This unusually big variation exacerbates the problem of estimating the genetic merit of an individual or group of individuals. An ILCA study (Peters, 1985) has shown that very large numbers of animals are required when comparing goat breeds in the tropics. For example, a group size of 700-1,000 animals is required to detect a difference of 5% (one-tailed test, â = 80%) if, as is often the case, the coefficient of variation lies between 35 to 40%. In these medium- to poor-nutritional environments, conventional genetic ranking procedures do not work effectively because of this large variation. McDowell (1983), analysing cattle records from a number of such environments, concluded that progeny testing, the standard orthodox method of sire evaluation, simply does not work in poor environments. What can be done in these situations? Ironically the lack of a breed infrastructure and this very large variation may be the key to future progress - progress through population screening.

Population Screening: In most breed improvement programmes, within-breed selection is limited to a rather small section of the population - the pedigree or stud breeders. This limits selection pressure, the most important determinant of genetic advance. In most livestock breeds in developing countries a pedigree breed structure does not exist and therefore selection can be considered at the level of the national herd. This may offer an opportunity to change a population genetically that has hitherto been ignored, since;

- most livestock populations in developing countries have not been subjected to controlled man-directed selection; and
- the large variation observed in these populations may in part be genetic, caused by major genes segregating at low frequencies.

Finally, it should not be forgotten that the major improvement in animal breeding, particularly affecting growth, body composition and conformation, made by livestock breeders in the last century, was achieved using very simple selection procedures. Breeders such as Robert Bakewell and the Colling brothers made dramatic genetic progress by 'screening' local animal populations for better animals as they subjectively assessed them. As a result, they developed the modern European breeds of livestock known today. They exploited the big variation in the then unselected livestock populations with which they worked. Bakewell is quoted as saying "Breed the best to the best and hope for the best". Since Bakewell's time, the genetics of animal production have become more fully understood and current problems can be approached with much greater scientific resolve and understanding.

To explain and hence attempt to capitalise on the unusually large variation that exists in many of the livestock populations in the developing countries, two hypotheses may be considered:

- That genetic variation in these unselected populations is larger than normal to the extent that the total variation is unusually large.

- That major genes with significant effects on production traits are segregating in these populations at low gene frequency.

Evidence to support the first hypothesis partially exists, in that reported estimates of the genetic parameters in livestock populations in developing countries are very similar, if not higher, to those reported for animal populations in the more advanced countries. There is no hard evidence to support the second hypothesis, since it has not been tested, but there is increasing evidence that the screening of national populations may identify major genes.

In recent years, at least five examples of major genes have been identified in sheep breeds (Australia, United Kingdom, Ireland, Iceland and Indonesia) and there are currently indications of a major gene controlling milk protein in a French goat breed (Timon, 1990). In any event, the screening of national populations for 'exceptional animals' makes good sense. At worst, it is a means of exercising maximum selection pressure within a population and, of course, it allows the possibility of identifying animals that are carriers of important major genes. In the difficult environmental conditions which predominate in many developing countries, the question of how to screen large populations in the absence of national recording schemes arises.

The only solution to this is to follow the same approach as the pioneer breeders of the last century. The initial screening should be done subjectively; in other words, the breeders are asked to identify 'exceptional animals', if any, in their herds or flocks. These 'exceptional animals' should then be recorded on the farm and compared with a random sample of contemporaries. If their performance approaches or exceeds twice the average level of the herd or flock, then those animals should be acquired, purchased or leased, for more controlled evaluation and breeding on a central test farm or research station. In this way, an open nucleus herd or flock can be established to act as a source of improver stock for the genetic improvement of the target population.

Screening Target: It is deliberately suggested that the selection target is pitched very high, at or near twice the flock average. Assuming a normal distribution for the trait in question and that animals are identified at the very upper end of the distribution, namely three standard deviations (o) above the mean (ì), it is easy to calculate the resulting selection differentials in terms of the variance or coefficient of variation (cv). If a selection differential ($i = i_2 - i_1$) is obtained that is equal to or greater than three standard deviations above the mean (i e" $3 o = 3i_1 \times cv$) then it can be calculated that 'exceptional animals' may be defined as being 100 to 200% above average, since the cv ranges from 30 to 70%.

If, on the other hand, a major gene or block of linked genes is segregating in the herd at low frequencies individual animals outside this range may be found. A simulation study of the possible effects of the Booroola gene in the Australian Merino and a specially selected (national screening) sheep flock in Ireland, has shown that coefficients of variation and repeatabilities in these flocks can be very high (60–70%) as a result of the expression of a major gene (Piper and Hanrahan, personal communication).

Population Screening - Genetic Rationale

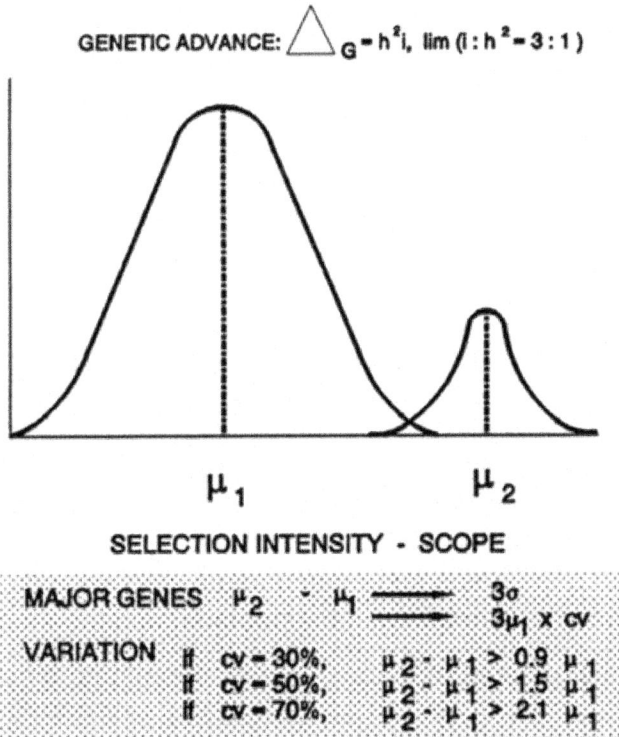

Figure 7: Selection Intensity Scope

Examples of successful screening have been reported in Ireland and Britain. In Ireland, the national sheep flock was screened for ewes displaying exceptional litter size, i.e., 3, 4 or more lambs/ewe (Timon 1964). The open nucleus flock which was then established had an average litter size of 2.3 lambs/ewe as compared to 1.3 in the national flock. Analysis of litter size distribution and repeatability estimates in this flock suggests the presence of a major gene or block of genes controlling ovulation rate. The Javanese Thin Tailed sheep (JTT) is also showing clear evidence of the existence of a major gene controlling ovulation rate (Bradford, Subandruja and Iniquez, 1986). This latter example is cited as evidence that major genes may and have been found in livestock populations living in harsh environmental conditions.

Screening of Awassi Ewes for Milk Yield: Following the logic outlined above, FAO initiated a preliminary programme to test the efficacy of genetic screening (GS) in conjunction with the establishment of Open Nucleus Breeding Scheme (GS/ONBS), as stage one in the genetic improvement of livestock breeds in developing countries. The initial programme was targeted at the genetic improvement of Awassi Sheep for milk production.

Pilot projects were initiated in Turkey, Syria and Jordan. In each project, the aim was to screen as many ewes in the national population as realistically possible; the screening was based on subjective assessment (based on interviewing the flock owner) followed by selective validation (measuring the milk yield of ewes claimed to be exceptional in controlled comparison with randomly selected contemporaries). Following this protocol, approximately 100,000 Awassi ewes were screened in Turkey, in a cooperative project with the University of Çukurova. Based on on-farm validation (two successive milk yield records), a total of 43 ewes were purchased and recorded in their following lactation alongside a control flock purchased (at random) from the screened population. The results, as shown in table, show that the screened animals outmilked the controls by nearly 40%. This initial response provides very encouraging, albeit tentative evidence that genetic screening for milk yield may point the way forward in the genetic improvement of Awassi sheep; the screening is being continued each year to build up an Open Nucleus Flock, which will ultimately provide improver rams for the industry. Similar screening projects aimed at improving body weight in Djallonke sheep in West Africa (The Gambia, Guinea) have been initiated.

Table 2: Genetic Screening Results - Turkey

	Nucleus		*Control*		%
Lactation	310	7.7	223	9.3	+39
Yield (kg) (range)	(254 -	469)	(97 -	360)	
Lactation	206	1.7	187	4.4	+10
Length (days) (range)	(159 -	224)	(95 -	222)	
Maximum	2.7	0.1	2.1	0.1	+29
Daily Yield range)	(2.1 -	4.2)	(1.0 -	4.3)	
No. of Animals	43		43		

Genetic Improvement Capacity

Very often development planners recommend a breed development programme, which, while technically justified, is without adequate emphasis on and evaluation of the capacity and commitment of the different 'actors' in the breed development chain.

Two issues are paramount here: that livestock breed development is a long-term problem requiring, at a minimum, a 20-year commitment; and that breed improvement cannot have any meaningful relevance if it does not compensate the farmer and the other participants in the breed improvement process over a sustainable timeframe and, to this end, research personnel, extension staff and support services (e.g. AI) must play strong and consistent roles to support effective breed development and farmer participation.

Further, long-term government policy and its impact on the development environment will also condition breed improvement possibilities. Consistency and coherence across all the areas in the breed improvement process, broadly determine genetic improvement capacity; all links in the chain are important to the final outcome.

Land use, Feed Resources and Integrated Crop-livestock Systems

The livestock feed resource base within any given land-use farming pattern determines animal productivity; conversely, the land-use pattern sets bounds on the role(s) of livestock within the farming system. The challenge to livestock development planning is to strike acceptable and sustainable balances between these sometimes conflicting interests. The diversity of issues, edaphic, biological, socioeconomic and cultural, that ultimately determine the accepted (traditional) farming system in any given region is very great and far beyond the scope of this paper. Only a few selected comments will be made here. Given the agroecological diversity of animal feed resources across the developing world, coupled with the variety of projects and past efforts to improve feed resources, the most that will be attempted is to discuss some general strategies on which livestock feed resources development and utilisation might be more effectively based.

Land Use

The still rapidly expanding human population in most of the developing world, which is projected to increase from 4 to 7 billion by 2025, continues to focus attention on land use, not only to meet the populations' staple food but also their living space requirements. The trends to Urbanisation across sub-Saharan Africa, West Asia/ North Africa, Asia and Latin America suggest that 54, 75, 56 and 84%, respectively, of the populations on these continents will live in cities by the year 2025 (Winrock, 1992).

Both of these fundamental demographic forces demand the attention of livestock planners now and towards the future. They dictate and emphasis on land-use efficiency in which the synergisms of soil, plant and animal interactions must be fully exploited. Coupled to this, concerns to conserve the natural resource base and to limit the emissions of carbon dioxide, methane and other environmentally damaging impacts, will further add to the complexities of livestock development planning.

Bioenergetic efficiency must become the keystone of livestock farming systems in the future. Inevitably, this means integrated crop-livestock systems throughout the tropics and subtropics. Certainly, in the low rainfall (<200 mm/year) rangelands, or in the high altitude alpine pastoral systems, crop livestock interactions will not be strongly evident, but in these environments avoidance of rangeland degradation through overstocking will be of paramount importance. Quite distinct but ultimately compatible short and long-term strategies are required.

Traditional feed resources: In the short term, the development of feed resources to raise ruminant livestock productivity throughout most of the developing world must be based on better use of traditional feeds. Generalisations are difficult to make across the diversity of production systems and their very different feed crops. However, the first step in all systems must be to tackle critical nutritional deficiencies, be they mineral-, protein- or energy-limiting.

In extremely difficult conditions, this may simply mean strategic supplementation of the diets of selected poorly thriving animals in order to reduce mortality. At another stage it will mean improving feed utilisation through: the treatment of by-product feeds (ammonia, urea); diet supplementation with balanced highenergy feeds, (urea/molasses blocks, etc.), coupled with the feeding of by-pass protein.

The underlying strategy being to build on the existing system and introduce simple and practical technologies to suit local conditions. Parallel to improvement in the feeding regime, due attention must be given to the major prevailing animal disease challenges. In the first instance, this will usually call for parasite control (internal and external parasites) coupled with vaccination against specific diseases as necessary.

New Feed Technologies: In the longer term, adequately supported by ongoing and future research, greater emphasis will need to be focused on the development and utilisation of high biomass feed resources (e.g. sugar cane, cassava) for monogastric and ruminant livestock, coupled with better exploitation of high protein forage trees. Expansion of livestock production in the tropics in the future will demand technologies that enable the ruminant to better utilise high fibre feeds, e.g. rumen flora manipulation to improve digestion of cellulose. Coupled with these changes, livestock planners in the future are likely to have to place much greater emphasis on multipurpose farming systems in which integrated livestock, food and tree crop production maximises photosynthetic capture to produce food (plant and animal), fibre, energy and fuelwood while maintaining soil fertility and overall sustainability within the system.

Examples of such integrated systems are already being developed to exploit the high biomass potential of the tropics (Preston and Murgueito, 1992). In other areas of the tropics, the more extensive exploitation of integrated rice, fish, azolla and livestock production is being researched (Mukherjee, 1992) in order to develop bioenergetically-efficient production systems to meet food needs in land-scarce countries such as China and South Asia.

The Development Force

Technical strategies for the advancement of animal agriculture constitute just one vector in the overall matrix of effects that determine the ultimate impact and sustainability of development effort. To place them in their proper context, it is useful to represent the development planning process as an algebraic function of the major determinant vectors.

For sustainable livestock development, these vectors will encompass: market demand for animal product or use, a_i; market/animal use channels and pricing arrangements, b_i; land and animal resource potentials and interactions, c_i; practical technical interventions and transfer impacts, d_i; natural resource sustainability and environmental impacts, e_i; and the major human components of the development process, f_i.

This last category ranges from farmer initiative and welfare to the efficiencies of the major government and private support service personnel involved in research, training, extension, market regulation, provision of capital and the formulation and implementation of government policy; these parameters of human endeavour might be conveniently termed the 'development force'. The challenge, inherent in the development of strategies for the sustainable advancement of animal agriculture, is to fully comprehend the direct (additive) and interactive (non-additive) effects of these different determinant vectors. They are represented in algebraic terms in the following equation, although it is fully understood that it is not possible to emperically formulate such an equation.

$$
SLD = f \; \; B_1 \begin{bmatrix} a_1 \\ a_2 \\ \cdot \\ \cdot \\ \cdot \\ a_n \end{bmatrix}, B_2 \begin{bmatrix} b_1 \\ b_2 \\ \cdot \\ \cdot \\ \cdot \\ b_n \end{bmatrix}, B_3 \begin{bmatrix} c_1 \\ c_2 \\ \cdot \\ \cdot \\ \cdot \\ c_n \end{bmatrix}, B_4 \begin{bmatrix} d_1 \\ d_2 \\ \cdot \\ \cdot \\ \cdot \\ d_n \end{bmatrix}, B_5 \begin{bmatrix} e_1 \\ e_2 \\ \cdot \\ \cdot \\ \cdot \\ e_n \end{bmatrix}, B_6 \begin{bmatrix} f_1 \\ f_2 \\ \cdot \\ \cdot \\ \cdot \\ f_n \end{bmatrix}
$$

Modelling and maximum likelihood iterative procedures can help to establish the relative importance (B) of each determinant vector on sustainable livestock development (SLD). Approached in this way, livestock development planning requires an interactive approach in which a number of development scenarios are identified; the interpretation of the outcomes will demand coordinated interdisciplinary analyses involving economists, sociologists, production scientists (land and animal) and environmental impact physics and chemistry. It is beyond the scope of this paper to critically analyse all the elements of each of the seven determinant vectors identified above. Selective brief comments will be made on two factors in the 'development force' vector in that they relate to issues that are most often ignored.

Farmer Initiative/Welfare: Many development programmes are based on the 'perceived' needs of the target farmers; the word 'perceived' is generally used by planners to indicate a full understanding of what the farmer needs! Farmer needs, like those of any other sector, are a function of tradition, education, and day-to-day survival demands normally influenced by the aspirations of the individuals, the extended family and the local community.

Increasingly, farming systems research and extension (FSR/E) approaches are attempting to put real perceptions on these needs; the livestock development planner must follow suit, essentially by 'standing in the farmers shoes' and perceiving development needs and objectives from that position.

Concerns for such matters as 'internal rate of return to investment', high output production systems and impacts of new technology will have to be transposed on such vital issues as investment risk, demand on work routines and family welfare since these are influenced by local tradition, culture and status in the local community. The development planning task is to marry these very different perceptions in an acceptable and phased development programme. The roles of the extension and other support services (e.g. credit) have a very important role to play in this reorientation/transition process.

Extension/Research Service Efficiency

Most developing countries, particularly in Africa, dramatically expanded their research and extension services over the past twenty years, basically based on World Bank investment loans (Pardey, Roseboom and Anderson, 1989). Yet it is difficult to single out many countries that can be deemed to have effective research and extension services to support the development of livestock production. Several International Service for National Agricultural Research (ISNAR) and other studies have identified and articulated many of the reasons why these services have not been successful in supporting change and sustainable advancement. Only two of the many reported constraints on research and extension services in developing countries, discussed in these studies, will be mentioned here; staff motivation and research-extension-support service links. These are often recognised, but are not always given adequate attention by development planners simply because they fall within the realm of national government policy and management capacity.

Staff Motivation: Staff motivation is certainly influenced by training, it can be nurtured or dissipated by management but, in the final analysis, it is heavily dependent on financial reward to the individual. Many research-, extension- or development-oriented projects have experienced great difficulty in harnessing effective commitment from national counterpart staff simply because the staff in question, albeit formally working full time in government service, find it necessary to have a second or sometime a third additional job/activity to earn enough money to support their families. It is unreasonable to accept that development objectives can be realised in such circumstances, although they may be very appropriate in every other way.

Support service efficacy is also heavily dependent on effective professional interaction between public service staff employed by different government departments or agencies. Sadly, it is not uncommon to find situations where apathy, jealousy and sometimes hostility discourage and limit the interactions, exchange of information and cooperation between the different sectors of government support services that are so essential to effective guidance and development of their farmer audiences. The sustainable development of animal agriculture, with its many complex interactions demands that these human initiative issues are fully considered in consort with the other elements in the sustainable development matrix.

Chapter 2

Sustainable Animal Agriculture: the Role of Economics in Recent Experience and Future Challenges

If a sharply rising population in the developing countries is to achieve higher real incomes and a better quality of life, agricultural output must rise more rapidly than population growth. Further, if this rise in agricultural output is to be sustained over time, the natural resources which provide the basis for such output must be preserved and new technologies offering higher productivity must be developed. Increasing livestock production is an important component of this process, both because developing country consumers are expected to spend an increasing share of their rising incomes on livestock products, and because taking advantage of favourable livestock-crop production interactions is one approach to a more efficient, sustainable agriculture.

Despite the generally favourable effect which the use of livestock has on agricultural resources, livestock are also a component of several unsustainable production systems. Among the most widely cited include:

- the overgrazing of arid and semi-arid lands, leading to range degradation and productivity decline,
- the destruction of humid rainforests for the establishment of pastures which degrade soils and quickly cease production,
- the pollution of watercourses from animal wastes resulting from intensive dairy and meat production, leading to reduced production and consumer welfare elsewhere,
- the production of methane gas as a result of ruminant livestock production contributing to the threat of a global "greenhouse" effect, and

- and the pollution of soils, watercourses and subterranean water supplies from the application of fertilizers, herbicides, and pesticides in the production of livestock feed grains.

Increased attention to the issue of sustainability in livestock production is important if these production systems are to be improved.

This paper is concerned with establishing a number of guidelines for optimising livestock production in developing or less developed countries (LDCs), consistent with achieving sustainable agricultural production. The paper focuses on economic factors (the author's specialisation) though it contains some thoughts on the role of political and social factors. The paper first provides an overview of recent achievements concerning livestock production in developing countries livestock production. The paper then briefly discusses how different types of livestock are used as capital goods to produce multiple livestock products, and how the total amount of livestock capital and its use is highly responsive to the economic incentives provided to producers. The paper then analyses several of the more prominent types of government interventions in livestock markets, discussing their probable effects on economic welfare. An emphasis is placed on the need for research. Finally, several problems involving sustainability in livestock systems are discussed.

Recent Experience with Livestock in the Developing Countries

Within numerous national and international agencies, including FAO and the World Bank, concern has been expressed with the progress, or lack thereof, recently observed in livestock development. What does the record show? Have efforts to encourage livestock development been successful and what are the prospects for the near future? What policy changes could improve the situation?

Consider first the evidence regarding the rate of growth of livestock output in LDCs (Less Developed Countries), both absolutely and relative to population growth. One comprehensive study notes that food produced from livestock grew more rapidly than population for most products in Asia, North Africa/Middle East, Sub-Saharan Africa, and Latin America between 1961-65 to 1973-77 (Sarma and Yeung, 1985). These results, with the exception for eggs, held for meat in all regions apart from Sub-Saharan Africa, but for milk only in Latin America. On average, the output growth rate for meat, milk, and eggs in developing regions were 2.9 percent, 2.5 percent, and 5.3 percent, respectively. Thus, per capita meat output grew about 0.5 percent per year during this period, though there were enormous variations across countries. Per capita milk output remained stagnant whilst per capita egg production rose rapidly. This record must be considered mediocre, especially since the consumption of livestock products generally rose at a more rapid rate, leading to rising imports or falling exports.

A more recent study confirmed this result for milk (Jarvis and Saha, forthcoming). Milk output in developing countries grew at a steadily declining rate from 1960 to 1988, falling from 2.7 to 2.1 percent per year. However, the rate of output growth varied significantly across geographical regions and from one sub-period to another which suggests that milk production has been responsive to changes in the incentives faced by farmers. For example, although the growth rate of milk output was low and declining in Sub-Saharan Africa and Latin America (±2 percent) during the last decade where per capita incomes were declining; it was high and increasing in East Asia (nearly 3.5 percent in the last decade). The regional growth in milk output is highly correlated with regional economic growth, suggesting that demand for milk has been an important determinant of milk production. Indeed, growth was highest where the resource base (natural pastures) were limited and vice- versa.

That milk output in the developed countries, which still accounts for 75% of world milk production, also declined during the last three decades. Nonetheless, subsidised exports from developed countries contributed to rising LDC milk imports and, in general, discouraged milk output.

Prices

Comparable data on real livestock prices are relatively difficult to obtain for a large number of countries, but it appears that, among meats in LDCs, beef prices have been relatively constant, again with wide variation across countries.

Domestic prices have often been tied, at least loosely, to international beef prices which determine the cost of beef imports or the value of beef exports. International beef prices have remained essentially constant during the last three decades (Jarvis 1986). Within Latin America, for example, beef prices have risen in a number of major producing countries as consumption as risen more rapidly than production, forcing countries from a position as net exporters to self-sufficiency. These governments have gradually reduced interventions which depressed beef prices.

In contrast to beef, poultry prices have fallen steadily throughout the world, mainly in response to increased production efficiency resulting from: improved genetics, balanced feeds, veterinary drugs, economies of scale and better management. Poultry consumption has risen rapidly in response to its declining real price and is becoming the meat of choice in many LDCs, particularly in urban areas.

Milk prices have fluctuated widely in response to the variation in the level of export subsidies from developed countries, especially the European Community (EC). However, many LDCs have imposed tariffs or quotas on powdered milk imports to protect domestic producers. India, South Korea, Brazil, Chile and Colombia are examples and milk production has responded positively. However, as will be argued below, in each of these cases, technical change and increased producer efficiency has contributed more to the growth in output than have higher prices *per se*. Indeed, milk output in each of these countries has been growing more rapidly than domestic consumption and, as each country has approached self sufficiency, governments have been reducing domestic prices. Future growth in the milk industry in these countries will depend on the growth of domestic milk demand and/ or each country's ability to export milk in an increasingly competitive world market.

Return on Investment

What has been the profitability of livestock investments? In a World Bank ex-post evaluation of 104 agricultural development projects which were wholly or partially devoted to livestock, roughly 60% of the economic rates of return (EER) exceeded 10%, while 34% were below 5% (The World Bank, 1985). The average return on livestock investments, weighted by the amount invested per project, exceeded 12%. Although these are only moderate returns relative to those estimated for other Bank investments, livestock projects have offered an acceptable return. More importantly, most of the projects which failed were in Africa, whose economic, technological and political context offered unusual challenges. More than two thirds of the projects in other regions performed satisfactorily.

Analysis of these projects suggested that the variation in their performance was determined by many factors (The World Bank, 1985).

Factors which had particularly strong effect included the availability or lack of:

- technological packages adequately adapted to existing farming systems,
- an economic context providing attractive producer incentives,
- an institutional capability for implementing the proposed project,
- qualified technical personnel to design and implement livestock development programs,
- a government commitment to livestock development, meaning a genuine willingness to establish a stable policy framework and commit the required public resources,
- political and economic stability, and
- the existence of clear property rights, which provide incentives for development and use of land and complementary resources.

Rather than suggest that there is something inherently difficult about livestock production, the World Bank's experience suggests that production depends strongly on the incentives faced by producers. These incentives, although strongly influenced by domestic income, population growth and by the international context within which countries trade, can be significantly affected by government policy.

Technical change is an essential component of any long term programme to increase livestock output, yet investment in livestock related research in developing countries has been low and the lack of appropriate technology has been an impediment the growth to livestock output.

In summary, livestock output has been growing, but ruminant production (beef and milk) have been rising more slowly than monogastric poultry and pork. The trend in beef and milk prices has been relatively constant, whilst of poultry and pork prices have steadily declined.

These figures suggest that technological change has been more rapid in the production of poultry and pork, allowing producers a favourable return while offering consumers a cheaper product. A higher level of technological change combined with protection of the resource base, is needed if ruminant output is to increase and provide consumers with more, cheaper and better products.

The uses of Livestock as Agricultural Capital

Livestock contribute in many ways to national welfare in developing countries. On average, livestock account for half of agricultural output when bothe their direct and indirect contributions are considered. Directly, livestock provide food and non-food products (hides and skins) amounting to about 20 percent of agricultural GDP. Indirectly, they contribute another 30 percent by supplying essential inputs to agricultural production.

Livestock convert crop residues, agricultural by-products and pastures on marginal lands (resources with limited alternative use) into a range of higher value products for subsistence and sale.

The integration of livestock into cropping, via draught power and manure, increases the area cultivated, improves the timeliness of agricultural operations and helps maintain soil structure and fertility. As economic development proceeds, the use of ruminant livestock by small farms will shift away from the current emphasis on traction towards beef and/or milk production (Jarvis 1982, 1988).

Because animal food products command high prices, they are usually consumed in greater amount by individuals having higher incomes, although, in some countries in Africa and Latin America they may be account for a high proportion of expenditure even in poorer households. However, even where the poor consume few livestock products, it is economically attractive for them to produce such products and exchange these for other foods which provide cheaper sources of energy and protein.

Both species and breeds vary in their capacity to produce different types of outputs and to utilise different types of inputs. Ruminants have the capacity to utilise low quality, bulky feeds such as pasture, crop and industrial by-products which have few alternative uses. Large ruminants (buffalo and cattle) also provide draught power and are by far the main source of milk.

Small ruminants (sheep and goats) are generally more prolific, produce wool and hair in addition to meat and milk and can prosper under poorer range conditions than large ruminants.

Small ruminants are also a more convenient household source of meat, barter and cash,. They are also more easily stolen. In many situations, large and small ruminants are complementary since they utilise different forage species. Diversification of disease risks is further reason for running them jointly.

Monogastrics (pigs and poultry) are utilised primarily for meat production, although hides, feathers, down and manure are also important products. The principal economic advantage of non-ruminants is their ability to convert high energy/protein feeds into meat at a more favourable ratio than ruminants. Such feeds are often expensive if they are demand in for direct human consumption. However, an increasing amount of high energy/protein feeds from crop by-products are now fed to non- ruminants and identification of cheap feed sources for monogastrics is an important aspect of livestock production in most developing countries.

In most countries there is a choice between two fundamental livestock production strategies a) the feeding of inexpensive, low-quality pasture resources to ruminants to produce meat, milk, wool, manure and draught power, and b) the feeding of high energy-high protein grains to non-ruminants for egg and meat production whenever the demand exceeds the amount which can be produced by ruminants from low-cost feed resources available. It is generally uneconomic to produce beef using feed grains, except where beef prices are unusually high, although the price of milk may justify such feeding.

Where pasture, forage or low-quality crop by-products are available (or can be economically increased), ruminants provide meat and milk at low cost. There is potential to increase pasture production in Latin America and so increase beef and milk production. A similar potential probably exists in much of Africa if trypanosomiasis can be controlled. In most other regions, pastures are limited and increased meat and milk production will have to come mainly from swine and poultry utilising grains and high-quality agro-industrial by-products. Such production will commonly take place in large, industrial-type enterprises close to urban centres.

Despite the rapid growth in demand for poultry, smallholders in developing countries can be expected to rely on ruminants as their primary livestock assets because they use more efficiently the locally available, low quality feeds, and that they provide a wider range of products, particularly draught and manure, crucial to their overall farming system. In meat production, smallholders can compete effectively with larger commercial enterprises only to the extent that they have access to low cost farm resources, especially feed and labour, which cannot economically be sold off-site. Such low cost resources usually result from the integration of agricultural and livestock activities. Many smallholders will find it profitable to maintain a small number of other species to utilise that available feed which ruminants do not utilise efficiently, to provide diversity to the family diet and to provide assets which can be liquidated in smaller amounts.

The Role of Government in Determining Livestock Production Incentives

Livestock are capital goods which are highly mobile and can be liquidated rapidly if economic incentives are unattractive. Livestock production can thus be strongly affected by government policy, both insofar as it affects: a) the prices which farmers face for their products and for the inputs they purchase, b) as it affects property rights (especially rights to land ownership and use), c) the development of new technologies, d) agricultural extension, e) the availability and terms of credit, f) animal health and sanitation, and g) infrastructure (e.g., roads, communications, and police and judicial services). Although it is impossible in a brief paper to cover the workings of all policies in all countries, a number of the most important policies affecting beef and milk in developing regions during the last two decades will be briefly analysed.

As a general rule, economists believe that governments should work to ensure that the price paid to domestic producers of, say, beef or milk is equal to the international value of these products, i.e., the FOB value of exports and the CIF value of imports in a local port, plus or minus domestic transport costs to the site of production or consumption, respectively. The same rule is valid for establishing the cost of productive inputs. When this rule is not followed, national economic welfare is usually diminished, though there are some important exceptions.

Historically, many governments (e.g. Latin America) have tried to reduce the retail prices of beef in order to benefit consumers. However, lower prices to consumers usually require lower prices to producers, which lead eventually to lower output. Thus, the effort to assist consumers via lower prices is usually successful only in the short run, and may actually harm consumers in the longer run as output declines. Such policies have been adopted at least once by nearly every Latin American country during the last two decades. Although initially all consumers might have benefitted through the availability of beef at a lower price, this benefit would soon be lost for many as consumption is reduced or higher black market prices. Of course, producers lose in both the short and the longer term due to the lower price received. That the analysis is similar if prices are imposed at the producer or wholesale level.

A similar effect will be achieved if a country which exports beef imposes a beef export tax, reducing the domestic price of beef below the international level. The relationship between import and export on meat supply and demand where the import price of beef is higher than the export price. Given that the intersection of domestic supply and demand occurs at a price lower than the export price of beef (P_e), it will be profitable for the country to export beef. If the government allows free trade to occur, the domestic producer and consumer price will be equal to P_e, domestic consumption and production will equal C_o and Q_o, respectively, and exports will equal X_o.

Suppose now that the government imposes a beef export tax equal to t, where t is some fraction of the export value. The domestic price will decline to $P_e(1-t)$, causing an increase in domestic consumption, a decline in domestic production, and a decline in exports. Moreover, it can be shown that economic welfare falls by an amount equal to the two shaded triangles. These losses occur because consumers are now artificially induced to spend more on beef and because producers are discouraged from producing beef, thereby, diverting their resources into other activities which the country produces less efficiently. In addition, the reduction in the domestic beef price causes a significant redistribution of domestic income. Beef consumers gain because the price of the product they purchase has declined; the aggregate gain for consumers is approximately indicated by the area abdc. In turn, producers lose because the price they receive for beef has also declined; their aggregate loss is approximately area aefc. The government earns tax revenue equal to the area tX_1. Because of the shift in income distribution, consumers may favour an export tax even though total national welfare is diminished by its effect.

It may be noted that some countries which have imposed a beef export tax have also neglected to develop new beef production technologies, perhaps in the belief that the beef export surplus will remain. Over time, however, beef consumption in such countries may grow more rapidly than production (shown by a greater rightward shift in the demand curve than the supply curve). If so, exports will decline. Once exports have been eliminated, the country will be simply self-sufficient and further increases in domestic demand will lead to a rising domestic price until the import price, P_m is reached. Subsequently, beef imports will begin.

Milk, like beef, is politically a highly sensitive product and has also frequently been subject to intervention. Such intervention has sought two distinctly different ends. In many countries, governments have concentrated on restraining prices. The effect has often been to depress output and decrease milk quality.

For example, in Colombia the government has imposed price controls at various stages of the production and marketing chain for pasteurised milk, but not for crude milk and processed products. The expressed intent was to reduce the cost of pasteurised milk, thereby benefitting poor consumers. The results have been somewhat different. Price controls on pasteurised milk have led to a) higher adulteration and lower milk quality, b) a higher percentage of milk marketed as raw milk with attendant health risks, and c) a lower price and higher consumption of processed milk products (cheese, yogurt, ice cream, etc.) which are consumed mostly by upper income groups, at the expense of a reduced supply of pasteurised milk, which is consumed more evenly by all income strata. A different situation is one in which the government has imposed an import tariff on milk to protect domestic producers. This policy was commonly used by developing countries during the mid 1980s when international milk prices were strongly depressed by subsidised exports from developed countries. A number of countries which had rising milk imports (eg. South Korea, India and Indonesia), made significant efforts to develop their milk industry to satisfy a growing national demand.

Where the intersection of domestic supply and demand lies above the import price of milk. If free trade in milk (powder) is allowed, the domestic price will be equal to P_m, domestic consumption and production are, respectively, C_o and Q_o, and milk imports equal M_o. If an import tariff of t is applied, the domestic price will rise to $P_m(1+t)$, inducing a decline in domestic milk consumption, a rise in production, and a decline in milk imports. It can again be shown that national welfare is reduced by the area of the two shaded triangles. The rise in domestic price induces consumers to shift their expenditure to other goods whose contribution to consumer's utility is somewhat less than that which would have been offered by milk at the initially lower price. Similarly, producers are induced to shift more resources into milk production, which provides less benefit than could be achieved had these resources been used to produce goods in which this country was innately more productive. That the import tariff which increases the domestic price again provides government revenue. It also increases producer incomes by an amount approximately equal to the area abdc, but reduces consumer welfare by an amount approximately equal to area aefc, which is larger than the gains achieved by producers and the government. Thus, milk producers are likely to favour an import tariff even if it decreases national welfare.

Although the conventional economic analysis laid out above indicates that the distortions introduced by a tariff on milk imports is likely to reduce domestic welfare, this analysis may be qualified in important ways. For example, if the international import price is temporarily low due to "dumping," but is expected to return quickly to a higher, normal level, a developing country may find it attractive to impose a countervailing import tariff to ensure that the domestic milk industry does not suffer irreversible harm during the period of low prices. An argument can be made that the gains from importing cheap milk during the interim period will be smaller than the long term losses inflicted on the dairy sector. Whether this is true depends on the relative costs and benefits of the alternative actions, which is an empirical issue likely to vary significantly from one country to another.

Similarly, if the country has potential to significantly increase its milk production efficiency in the long run, a transitory import tariff may permit, if combined with other actions like a strong research program, the realisation of such potential. This "infant industry" argument for a tariff is often advanced for the manufacturing sector. A problem, however, is that the alleged increased efficiencies sought are often not achieved and the short run tariff becomes a long run tariff, protected by vested interests. If so, although producers gain from a higher domestic price, the loss by domestic consumers will be larger. In a number of countries, including India and Indonesia, domestic dairy development has been encouraged by significant import tariffs and/or the receipt of concessionary milk imports which were sold by the government at market prices. The revenues thus obtained being used to subsidise the development of the domestic dairy sector

(Alderman). Although such efforts have achieved significant improvements in the quantity and quality of milk produced, in each case, consumers have had to pay higher prices of milk than they would have had to pay had milk been imported at cheaper prices. Thus, if cheap milk was desired for consumers, the policy of applying tariffs has achieved the reverse, at least in the short run. However, since higher income consumers account for the bulk of milk consumption, such policies might have distributional justification in some cases.

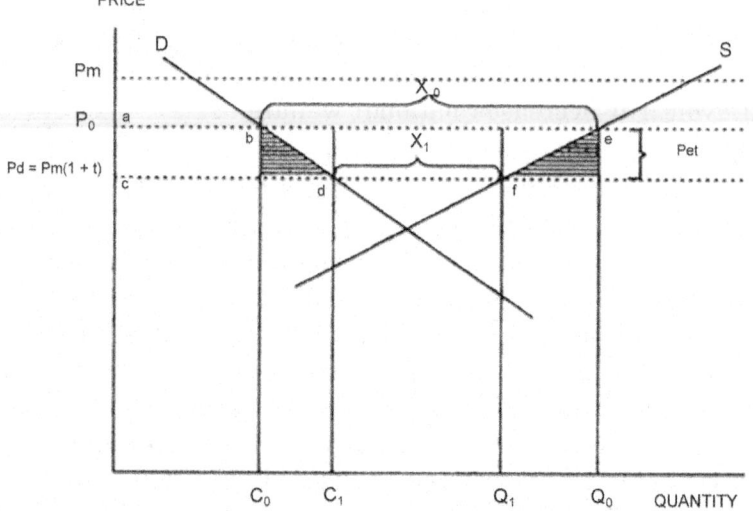

Figure 1: The Welfare Effects of a Beef Export Tax

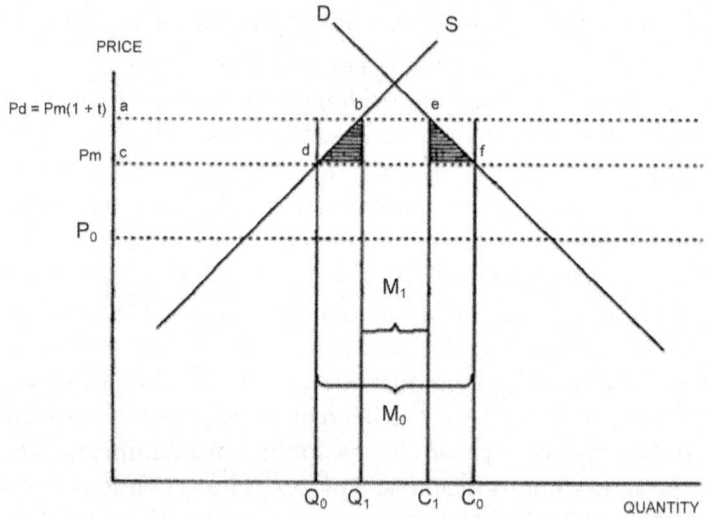

Figure 2: The Welfare Effects of a Milk Import Tariff

Price intervention via export taxes, import tariffs, and fixed prices are common in developing countries. This paper has attempted to show that they can cause significant economic damage, even where they increase livestock output. The appropriate policy goal ought to be to optimise total economic output, not the output of any specific product - even livestock!

Several other types of economic interventions warrant mention. One is exchange rate overvaluation, which occurs when the domestic currency purchases more foreign currency than it would at an "equilibrium" in the balance of payments. Exchange rate overvaluation generally occurs as the result of economic development policies which restrict imports to encourage industrial development. The reduction of imports reduces the demand for foreign currency, which in turn leads to a stronger domestic currency. The stronger domestic currency directly reduces the domestic value of all traded commodities. For exported goods, for example, it has an effect similar to that of an export tax. It will do the same for imported products if these are not protected by tariffs.

Empirical studies of policies implemented in many countries suggest that the indirect effects of currency overvaluation have probably been more damaging to the livestock sector than have the direct interventions like price fixing and export taxes, which are more visible (Krueger, Schiff and Valdes, 1988). Rarely have agricultural sector spokespersons fully appreciated the damaging effects of currency overvaluation. Even when they have, however, they have usually had little influence on such policies. Although agricultural interests usually influence the policies which directly affect the agricultural sector, the influence of such interests on macroeconomic policies, i.e., the exchange rate, is usually limited.

Government have occasionally sought to prohibit certain types of beef slaughter. For example, in Colombia in the mid-1960s and in Chile in the early 1970s, slaughter of heifers and cows that were deemed still "productive" was prohibited in an attempt to force expansion of the breeding herd and thereby output. This measure led to increased clandestine slaughter of females. Similarly, in Indonesia, the slaughter of younger bulls was prohibited with the intent of building up the livestock herd. It can be shown, however, that the constraints imposed on producers by these prohibitions reduced overall livestock profitability by impeding the producer from taking what appeared the most profitable action (Jarvis 1982). Farmers generally perceive better the appropriate use of livestock capital than do governments. In each of these cases, when farmers slaughter animals they do not generally cease to own livestock assets, but simply shift from ownership of an older less efficient animal to a younger more efficient animal. The government's effort to dictate the appropriate use of animals thus doubtlessly led to a long-run decline in the livestock herd and in output rather than the increases sought.

In most developing countries, industrial protection policies (import tariffs and quotas) has increased the cost of inputs important to livestock production, such as machinery and equipment, veterinary supplies, minerals supplements, fertilizer, and fencing materials are usually high relative to costs in developed countries. The lack of competition among importers of many brand name products is also a factor in high input costs. Such costs inhibit the development and adoption of improved livestock technology.

Given the tendency of government market intervention to cause harm, most economists argue that governments should minimise such interventions. However, there are other non-market interventions, like those in public health, veterinary services, research and extension and the provision of infrastructure, which governments must undertake.

Livestock Research, Infrastructure, and other Government Interventions

It has become increasingly clear that achieving higher rates of livestock output associated with higher incomes for producers and lower prices and higher quality products for consumers, depends on obtaining improved technology. Probably no aspect of government policy is as important as research for the future of livestock production in developing countries. In several countries where efforts have been made to develop a domestic dairy industry, via import protection, the increase in output has probably come more from technological diffusion than from the stimulus of higher prices, *per se*. For example, in India, the increasing use of crop by-products as feed concentrates, along with the introduction of crossbred dairy cattle having higher potential yields, has resulted in a rapid increase in milk output per animal. Similarly in Colombia and Brazil, the spread of pasture raised crossbred dual purpose cattle has led to rising output through a combination of increased herd size and higher output per animal. In each case, the output expansion has occurred while prices have been roughly constant or declining. Further, as domestic output has risen, reaching self-sufficiency levels, policy makers have been confronted with the possible need to export milk surpluses at world prices which are substantially lower than either the protected domestic price or, even, the milk import price.

Even when basic technology is available in developed countries, a strong domestic research capability is essential to identify and adapt promising technologies to local conditions. Unfortunately, most livestock research and development institutions are weak and a major effort is needed to strengthen them, especially, the number of qualified staff.

An institutional environment which allows farmers to a) market their produce for higher prices, b) obtain inputs at lower cost, and c) obtain information necessary for decision marking more rapidly and effectively is critical to livestock development. Infrastructure is especially important where timely access to inputs is essential and where livestock product must be marketed quickly. Milk production is such an example and depends heavily on an efficient the establishment of a marketing and distribution system, including milk collection, transport, and processing. The diffusion of new technologies to farmers, e.g. artificial insemination and forage technologies, has often been achieved through the same systems. Such efforts have been carried out efficiently by private firms and cooperatives, though rarely by public firms.

Livestock diseases impose significant economic losses and can present a threat to human health. Such losses and threats can be economically reduced by improved animal health programmes. However, programmes such as vaccination, sanitation and inspection programs require a deep commitment to implementation from producers and governments and often require regional cooperation.

The Effect of Social and Political Factors

Social factors can influence the use of livestock such influences are often obvious, for example, religious objections to the consumption of pork in Muslim countries and beef in Hindu countries. In other cases, however, the effects of social influences are less pronounced though still important. For example, the attribution of prestige to ownership of cattle may increase the cattle's total value but lead to other tangible products like meat and milk. In general, such effects seem relatively small in magnitude.

In much of Asia cattle, utilising locally available residues, are used mainly for animal traction on small farms. In India, for example, cattle are fed principally on wheat and rice straws which account for approximately half the plant's total gross energy. This energy would be largely wasted if it were not consumed by ruminants and converted into traction for ploughing and manure.

Due to religious beliefs beef in India is of little value (beef is consumed by the Moslem minority in some areas). Nonetheless, farmers require some means to cultivate and most farms are too small to justify mechanisation and produce forage. The value from draught and manure must therefore be sufficient to justify maintaining cattle and this in fact the case where livestock are kept. Incorporation of manure into fields and garden areas, to improve soil fertility and structure, is often essential to maintaining agricultural production. It can be shown that the social constraint on the eating of beef, by diminishing the value of beef which is a necessary by-product of milk, traction, and manure production, makes cattle a less profitable investment. Further, with no incentive to slaughter the animal, the tendency is for producers to simply abandon their animals when their economic life animals has ended. Such animals are a general nuisance and harm other farmers who need to protect their crops and from consuming valuable common grazing.

Chapter 3

The Sustainability of Livestock Production Systems

Livestock are usually helpful in sustaining agricultural production. However, there are cases where livestock development has had disastrous environmental consequences. For example, clearing of the tropical forests in Central America and the Amazon during the last two decades, these developments has been sharply criticized for their ecological and sociological damage.

Most criticism has focused on a) the destruction of irreplaceable genetic materials, b) tendency for pasture to rapidly diminish in productivity because of loss of soil fertility, leaving the fragile soils vulnerable to compaction and erosion, c) the displacement of indigenous peoples and small farmers by land speculators who have used cattle ranching as a mechanism for obtaining and controlling large tracts of land, and d) the threat to the environment from destruction of oxygen producing trees.

Livestock development, *per se,* in most of the Amazon basin is not very profitable at current prices. Nonetheless, government incentives in Brazil have affected livestock development and, more dramatically, Amazon settlement and deforestation (Binswanger). Income tax credits and subsidised interest rates on loans for livestock development, along with grants of land on favourable terms to individuals engaged in livestock development, have given substantial private incentives for livestock development in rainforest areas.

This is one of the most dramatic examples of a case where government policy is the primary cause of an unsustainable agricultural system. Although some rainforest destruction would remain even if government policies were fully neutral, due to the pressure of spontaneous colonisation by poor farmers but the areas affected would be much smaller. The damage caused by such settlements is a more difficult problem. Achieving a sustainable system in such situations will require development of either improved technologies or, more likely, the exclusion of settlers.

Developed countries contribute to environmental degradation to a far greater extent than does either Brazil or other developing countries. However, that others have and are destroying irreplaceable assets is poor justification for continuing with equally bad policies in the developing countries. The intent should be to make the best use possible of the available resources, in all regions.

Overgrazing on semi-arid and arid lands, leading to range degradation, is another case of an unsustainable livestock system. Pastoralism is a practice that utilises extensive rangelands where rainfall is low and highly variable, making settled agriculture and/ or livestock production extremely risky. The principal production risk for a specific area is that no rain will fall. However, cattle can be herded to areas in which rain has fallen and pastures are available.

The need to access a large area of land in order to ensure sufficient pasture is an important reason for the evolution of "common" range systems. In such systems, a group of pastoralists share land, with all being able to move about with their herds in search of the best forage. If the system is to work well, pastoralists must have a well defined membership group with clear (albeit sometimes complex) rules of access to pasture and water.

If group membership and/or the rules of access become unclear or ineffective, particularly as when population or economic pressures encourage greater use of the range, the system may tend toward an "open" access system in which no limits are placed on the number of herders (and animals) using the land. In this case, the economic value of pasture is likely to be severely diminished or lost altogether (Jarvis 1984).

In most areas, pastoralism is probably more productive in terms of the value of total output of beef and milk per hectare of land than is cattle ranching. The pastoralists utilise much more labour and extract a larger number of joint products for direct use, especially milk. Beef is produced from cull animals, both steers and cows, but accounts for a relatively small proportion of output. Pastoralists often barter milk and beef with agriculturalists for grain, which is a cheaper source of energy. Manure is used for fuel, and is also left on the fields of agriculturalists during seasonal migrations into settled areas. Agriculturalists sometimes pay herders to graze their animals overnight on their fields.

In systems in which land is communally owned, livestock ownership provides usufruct rights to land which are otherwise lost. Mechanisms are needed in such systems to ensure that all individuals having grazing access also have livestock. In pastoralists systems, livestock ownership traditionally belonged to kinship groups which used force to maintain their hegemony over a particular region. Complex societal rules and livestock exchanges existed within such groups to ensure that individuals who lost their animals to disaster, such as drought or disease, could reconstitute their herd. Such mechanisms have been breaking down in recent years, largely because pastoralist populations have gradually expanded while rangelands have been lost to the spread of sedentary agriculture. Under these conditions, the average herd has been shrinking whilst the aggregate number of animals grazed has been growing.

Traditional mechanisms have proved insufficient to reconstitute the herds of many individuals following disaster. Wealthy individuals, frequently located in urban areas and able to better diversify risks through other economic activities, are accumulating animals and hiring others to herd them. Gradually, as a higher proportion of total herds are owned by such individuals who seek a more marketable output, greater emphasis is being placed on beef production.

Under such pressure, there is fear that overgrazing is increasing. It appears, nonetheless, that the main effect of overgrazing has been an increase in the periodic herd losses suffered from drought, rather than a decrease in range quality.

The range generally seems to have substantial resiliency, recovering more rapidly than the herd. However, such a system is truly sustainable, except perhaps at a low average and highly variable yield. A greater problem is the lack of incentives created by the common range system for the development of any productivity increasing technologies. Such technologies would be privately unprofitable within a situation where animal nutrition is not under the herder's control. Without a shift toward greater control of land, there is little possibility of increasing ruminant livestock output in this area.

The primary problem faced by common ranges is the inequities which are likely to be created as common lands are converted into lands with an increasing degree of private control. However real, important and difficult these issues are, it is not an adequate justification for retaining the current system.

Historically, pastoralist groups have fought for specific areas, thus deciding "property rights." In recent years, the increased value of livestock output has led to the gradual emergence of more private land rights in countries as diverse as Somalia and Botswana. Kenya has successfully privatized its land with beneficial results. Clarification of such land rights is the primary factor needed to reduce overgrazing. However, that will not in itself reduce the increasing pressures on land which are stemming from higher population and which lead to deforestation and cultivation of marginal land. Again, the best solution seems to require the development of improved technologies which are more productive and sustainable combined with the education of producers to use them.

Feed availability, disease, climate, social and political forces, as well as economic incentives all influence the pattern of livestock use which emerge in developing countries. Thus, one of the main responsibilities of policy makers is to ensure that economic markets work well so that livestock producers receive appropriate signals regarding resource allocation. Similarly, the sensitivity of the mix of outputs to economic and technical factors indicates that, when formulating livestock development strategies, it is important to have a clear understanding both of the various production constraints and also of the demand the for different products.

There are cases where government intervention in markets is justified i.e. the protection of domestic producers from international dumping or to initiate long term industry development. Unfortunately, development often does not always occur even when the government has intervened to obtain higher livestock prices.

Sometimes this is due to the intervention appears short term and the resulting uncertainty mitigates against development which were in fact sought. More generally, the government does not make the concerted effort through related research to develop the technologies required to increase productivity, or the requisite political stability or land rights are not present. As a general, though not precise, rule of thumb, livestock development in most countries will make the greatest contribution to domestic welfare if the government establishes free markets for livestock products and inputs and strives to develop complementary research, infrastructure and animal health programs. Research, education, and non-distorted market signals seem generally to be the best guarantee of achieving sustainable livestock production.

Investment for Sustainable Livestock Development in Developing Countries

Bank livestock lending peaked in the 6 year period 1974-79 when an annual average of seven free standing livestock projects and 19 projects with a livestock component were approved; costs including the cost of livestock components amounted to US$9.47 billion (1989 US dollars) and 5.15 billion (U.S.$ current). Livestock lending declined in the period 1980-85 to a yearly average of about two free-standing projects and 17 component projects; costs amounted to 5.77 billion (1989 US dollars).

Over the five year period 1986-90 the aggregate cost of Bank-assisted livestock projects and components amounted to US$3.54 billion (1989 US dollars). Average annual lending declined in real terms from US$1,578 billion for period 74–79 to US$962 million for period 1980 to 85 and US$708 million for period 1986–90. This means that average lending for the last two periods was only 61% and 45% of the earlier period (1974–79) in real terms.

For the first time no free standing livestock project was approved in FY90 but 11 projects were approved with a livestock component. The absence of a free standing project in FY90 is considered as a one-year anomaly and it is expected that the number of livestock projects per year will stabilise at about 2 free-standing and 12–15 component projects.

The reduction in livestock lending as a proportion of the total lending for agriculture parallels, albeit to a greater extent, agriculture's decline as a percentage of the Bank's lending operations; from 30.1% in 1980 to 16.3% in 1989 and 17.3% in 1990. The character of lending operations changed substantially over the last decade with greater emphasis being placed on lending instruments other than specific investment loans. IBRD and IDA lending by loan category is given below for FY90.

Table 1: Lending FY90

Loan Category	US$Million	%
Specific investment	19,179.9	49.15
Sector investment	3,957.2	19.11
Financial intermediary	879.0	4.24
Sector adjustment	2,543.6	12.28
Program lending and Structural adjustment	1,434.0	6.92
Debt reduction	1,460.0	7.05
Technical assistance	203.0	0.98
Emergency reconstruction	54.0	0.26
	20,710.7	100.00

A decade ago Bank lending was dominated by specific investment loans but the picture has changed drastically with the increased emphasis on other types of loans, especially sector investment loans, sector adjustment loans, program lending and structural adjustment loans.

Review of Bank Experience with Livestock Lending

The reduction in Bank lending for livestock was caused by the relatively poor performance of livestock projects as indicated by a number of reports and papers which reviewed lending for livestock development. The most important and comprehensive of these reviews- "The Smallholder Dimension of Livestock (1985)" - was undertaken by the Bank's Operations Evaluation Department (OED) which is entrusted, by the Bank's Board of Directors, with authority and responsibility to undertake an independent review and report on the performance of all Bank lending operations.

In addition to the OED report the Bank's experience with Dairy Development was reviewed in 1982 (AGR Technical Note No. 6). The Bank's experience in Dry Tropical Africa was reviewed in 1981 by Mr. Stephen Sandford (Consultant) who produced a report for internal use. These three reports provide a good independent assessment of livestock lending including problems and issues and suggested lessons. Four World Bank reports which dealt with special aspect of livestock, e.g., veterinary, dairying and integrated crop- livestock are not dealt with in this paper although they provide valuable insights on livestock lending and the sustainability of development efforts. Salient points from these papers are, however, discussed later in this paper.

OED Report

The OED report was based on a review of 124 audited projects and 206 ongoing projects which comprised the Bank's livestock portfolio at the end of 1983. Of the 330 total, 91 were livestock projects and 239 livestock component projects. Of the 124 audited projects, 52 were livestock projects and 76 were component projects.

From modest beginnings in 1959 in Uruguay, through late 1983 early 1984 when the OED study commenced, the Bank provided some US$11.7 billion (in constant 1983 dollars) for livestock development of which almost US$6.1 billion (52%) was targeted to smallholders. The livestock sub-sector was thus significant in the Bank's lending portfolio and smallholder livestock lending was an important part.

The rate of livestock lending increased through the 60's and 70's to peak in 1979 and thereafter to decline. Smallholder lending followed a similar pattern but the percentage of total livestock investments for smallholder development showed a steady increase from the late 60's to the present and over the 70's accounted for roughly two-thirds of all Bank livestock lending. This was in response to Bank management's increasing sensitivity in that period to equity considerations which affected all agricultural sub-sectors.

There were according to the OED report regional differences in project components and their design, in species/product emphasis and in target group, depending on ecological, socioeconomic, cultural and traditional factors. Credit and livestock purchase were major project components in most regions. Nearly two-thirds of all audited projects included such components. About half of all projects included components for development of on-farm infrastructure, pasture improvement, and fodder development. Projects frequently included the above activities as multiple components. About 40% included a technical assistance and/or farmer training component. Other components occurred less frequently. In general, the design of smallholder projects was similar to that found for the whole set of livestock-related projects. The ongoing projects also had larger scope than the audited projects and nearly all of the components appeared with greater frequency suggesting that each project, on average, included a larger number of livestock-related components.

In terms of species emphasis, the OED report found that investments in cattle development (beef, dairy, and dairy-beef, in that order) accounted for about two-thirds of component activity in audited projects and an even higher proportion of total funds. Substantially less attention was paid to other species, viz. sheep, poultry, and swine, each about 10%; goats, about 5%; and miscellaneous species (cameloids, rabbits, bees, etc.), about 2%.

Cattle development was also emphasized in ongoing projects, but its relative importance decreased to less than one-half of the total number of interventions. Thus, while cattle development continues to be important in ongoing projects, other species are now increasingly emphasized. This emphasis on other species reflects increased efforts to support and improve existing smallholder farming systems. In such systems there is also growing emphasis on integrating livestock with agriculture, to their mutual benefit.

All regions participated in the Bank's livestock development efforts according to the OED report. The number of livestock-related projects (audited and ongoing) at that time was spread fairly evenly by region, but the bulk (three fourths) of all livestock investments were in Europe, Middle East and North Africa Region (EMENA) and Latin America and the Caribbean Region (LAC). Smallholders received the highest proportion of livestock investments in Western Africa Region (WA), South Asia Region (SA), EMENA, and East Asia and Pacific Region (EAP) in audited projects, and in Eastern and Southern Africa Region (ESA), EAP, SA, and WA in ongoing projects.

Project Performance

The average ERR of all audited livestock activities Bank-wide was 11% (OED report). The average ERR of 46 livestock projects was 7.2% and that of 58 component projects, 14%. The lower average figure for livestock projects appeared to be due to the low returns from smallholder livestock projects (average ERR -0.3%) and large holder livestock projects (average ERR 6.2%), as mixed smallholder/largeholder livestock projects performed satisfactorily on average (ERR 11.6%). Smallholder component projects (average ERR 10.7%), largeholder component projects (average ERR 18.9%) also performed generally satisfactorily.

While the results suggested that livestock investments made as a component of a diversified project were most successful than as a part of a straight livestock project, there was insufficient data to support such a conclusion statistically because separate ERRs were rarely available for individual components of multi-component projects, particularly if the individual components were relatively small, as they often were for livestock investments. It was stated that additional information was needed on the performance of livestock investments within livestock component projects, particularly as such investments comprised an increasing proportion of the total ongoing livestock portfolio at the time of the OED Study.

By region the OED Report showed that, EMENA, LAC and Southeast Asia had the highest number of total projects with ERRs exceeding 10% (77%, 68% and 67%, respectively) and ESA, EAP and WA had the highest number below 10% (74%, 56% and 50%, respectively). Some 14 of 22 projects (64%) with negative ERRs were in ESA and WA.

By project type, livestock projects performed particularly poorly on average in both African regions and, to a lesser extent, in East Asia and the Pacific Region (EAP). Component projects also performed unsatisfactorily on average in Eastern and Southern Africa (ESA), but performed satisfactorily overall in all other regions, particularly in EAP which had the highest average ERR (33.7%) Bank-wide.

Ex-ante appraisal projections were observed to be much more optimistic than *ex-port* ERRs. *Ex-ante* ERRs on 46 livestock projects were some 200% higher than ex-post ERRs and on 58 component projects they were some 70% higher. Only one in six of all projects has ERRs at completion equal to or greater than appraisal ERRs.

Two conclusions emerged from the OED Study. First, a large number of livestock investments were successful, particularly in the regions where lending was highest, and their success should not be obscured by the existence of problem projects, particularly those in ESA and WA. Second, the substantial variation in project performance suggested a need for improved appraisal methods, especially greater attention to the production coefficients adopted, the benefit stream projected, the project time frame and risk analysis.

The principal factors affecting project outcome identified in the OED study were the availability or lack of:

- technological packages adequately adapted to existing farming systems;
- an economic context providing attractive producer incentives;
- the institutional capability for implementing the proposed project;

- qualified technical personnel;
- a government commitment to livestock development and/or smallholders;
- political and economic stability;
- clear property rights for lands to be developed;
- functioning producer organisations - particularly where group action is needed;
- a project design which realistically takes into account country strengths and weaknesses; and
- firm, consistent, but flexible supervision of implementation.

The OED report stressed that these factors were similar to those which cause problems in projects in other agricultural sub-sectors. An effort was made to identify factors specific to livestock. The risk in livestock projects appeared most closely linked to the inadequacy of applied livestock-related research in most countries, the lack of technical personnel, the weakness of livestock-related institutions and their lack of integration with agricultural institutions, the greater importance of land tenure issues, and the failure of some projects to recognise fully the inadequacy of the "base" on which projects had to build. The tendency to proceed too rapidly in terms of physical implementation, without sufficient technological, institutional and staff development, stood out.

The study also pointed out that the performance of livestock projects must also be measured in dimensions other than the simple ERRs. Many livestock projects were pioneering efforts, involving new relatively untested technologies, requiring institutional strengthening, staff development, and livestock policy formulation. Benefits achieved through improved institutions, staff development, and "learning by doing" were not revealed in the ERRs. Nonetheless, there was a strong correlation between the project ERRs and the audit reports assessments of achievements in these other areas - poor economic performance has often been accompanied by poor institutional development and the like rather than offset by improvements therein. Indeed, poor performance in these areas was a major cause of low ERRs.

The smallholder livestock and component projects examined by the OED Study fared worse than livestock projects (taken as a group). Their average *ex-ante* appraisal ERR was equally as high as other livestock projects, but their *ex-post* ERR was even lower.

The OED study concluded that the widely-held perception that the Bank's livestock development efforts have been unsatisfactory accounted, in part, for the steep decline in livestock lending since 1980.

The study found that the performance of Bank livestock was highly variable, ranging from very satisfactory to very unsatisfactory, but was satisfactory more frequently than not. It also appeared that considerable learning had taken place regarding the design and implementation of livestock projects, and it was concluded that future projects should perform better.

Nonetheless, it also concluded that the evidence suggested that livestock projects overall may be more difficult than other agricultural sub-sector projects, and that livestock assistance, especially to smallholders, would probably require greater design, implementation, and supervision inputs than they had received up to that time. The study also concluded that the Bank should continue and probably increase its support for livestock development, especially to smallholders, given the high potential for raising their incomes and living standards.

It pointed out that livestock were a key element in raising farm productivity and it was difficult to conceive of sustained increases especially in smallholder agriculture in most areas in the world without attention to livestock development; demand for livestock products was increasing rapidly in most developing regions, and livestock investments were expected to be increasingly economically attractive. The study highlighted the point that livestock project design initially placed heavy emphasis on meat and milk and largely ignored other outputs such as traction and manure. This approach was strongly influenced by livestock systems in developed countries, and by an emphasis on larger commercial producers.

The move toward smallholder livestock development had encouraged a shift toward more diversified use of livestock and the integration of livestock and agricultural production activities. It emphasized the need for this shift to be more fully reflected in project design, e.g., greater cognizance should be taken of joint livestock outputs when assessing the demand for and the benefits of livestock production, and greater effort should be made to coordinate the efforts of livestock and agricultural development agencies.

Milk was considered to merit greater emphasis relative to meat production. The primary need was seen as the organisation of marketing, processing and distribution facilities, especially in regions where milk production was dominated by smallholders. Small-scale dairying was seen as undoubtedly one of the most promising avenues for future Bank lending.

It concluded that livestock development efforts suffered from a "cattle bias". Additional emphasis should be placed on swine and poultry, small ruminants and other animals, especially through research, technical services, and market development.

The study made special mention of livestock-related projects in the African regions because of the difficulties which were experienced there. It pointed out that many of the countries in ESA and WA were newly independent with ill-defined policies and priorities, limited infrastructure, weak skilled manpower and material resources, widespread poverty, and a high prevalence of drought and pestilence. Many governments were overly centralised and urban oriented.

Not only livestock investments fared poorly in such countries, but indeed all agricultural-related activities. Nevertheless, in several countries, livestock development appeared crucial to overall economic development and, in others, it would have a high positive impact. It emphasized that a substantial amount had been learned regarding livestock development, and there was evidence that where such lessons had been applied in ongoing projects the situation was improving.

Finally, it was evident from the study that smallholder livestock projects performed unsatisfactorily overall. The review indicates that efforts were often made to develop individual projects which were innovative, but these were also often ambitious in scope and size, were generally weak technically, and were implemented in a largely unfavourable economic climate in countries where government sometimes showed limited sensitivity to smallholder development potential and needs.

Target groups sometimes failed to involve themselves in project design and implementation out of misunderstandings or from distrust of government intentions, and were other times excluded either for paternalistic or political reasons. The Bank may have been too ready to finance such ambitious projects, particularly where institutional support was weak, where land tenure problems were apparent, and where government commitment was questionable. In a number of instances, an exploratory pilot phase would have been more appropriate instead of a large, demanding and high-risk effort. The study mentioned that a number of operational staff hold the view that pressures of the lending program were a contributing factor in this context.

Review of Bank-financed Dairy Projects

In 1982 the Bank prepared a comprehensive review of Bank/IDA-financed Dairy Projects (AGR Technical Note No. 6). The paper lists and reviews 75 projects which were either exclusively for dairying or had a dairy component. The total cost of these projects amounted to US$6,533 millions, the dairy components amounted to US$1,034 million and the loan/credit amounted to US$1,999 million.

Latin America (LAC) Dairy Projects

The study concluded from a review of 30 dairy programs in LAC that the projects "successfully attained production targets, improved the institutional support structure and contributed towards the establishment of effective livestock credit systems".

A feature of the region was the traditional preference for raw milk (which is boiled before use) and this enabled smaller producers to dispose of surplus milk directly to consumers without concern for an accessible processing facility. Low priced imports were a constant threat but it is concluded that "the livestock industry throughout the area had adjusted to the low price import option by developing a dual purpose production system, by utilising natural pastures, by upgrading cattle and by increasing the carrying capacity of the land". Project officers considered that marketing patterns did not need changing and processing facilities, especially for pasteurised milk, were adequate although increased investments would be required as dairying expanded. The report draws attention to the deleterious effect of low cost imported milk products on investment and production in many LAC countries but concludes "in the LAC area it is doubtful that there is a country or group of countries which could be described as marginal for dairy production in terms of the possible financial advantages of imported milk projects".

Europe, Middle East and North African (EMENA) Dairy Projects

Fifteen dairy or projects with dairy components were assisted in six countries (Morocco, Turkey, Yugoslavia, Ireland, Spain and Romania). Total project costs were US$2.47 billion and dairy development components amounted to US$778.01 million and Bank loans amounted to US$597 million. Projects are judged to have performed satisfactorily in all cases with the exception of large public sector farms in Yugoslavia. They encountered serious management, overstaffing and price problems. Yugoslavia changed its policies as a result of this negative experience and subsequent projects supported smallholders only.

Dairy programs in virtually all countries emphasized pasture, forage and feed production. Successful dairy cattle importations were a feature in the case of Turkey and Morocco (Friesian and Brown Swiss). Another feature of these latter countries is the prevalence of raw milk consumption which enables producers to sell directly to consumers and traders without incurring processing costs. The Rumanian projects main objective was to improve yields and labour productivity by modernising existing large public sector units (involving feeding, housing and milking). With the exception of Ireland, which exported most of its milk as manufactured products under EEC arrangements, production in the EMENA projects was almost entirely for home consumption to meet increasing consumer demand.

The report concludes: "Assessment of overall performance of EMENA dairy development projects provides evidence among most projects of measurable success in achieving project goals and governments' major objectives. The diversity of project design grew out of the need to develop approaches best suited for particular social, economic, and ecological conditions country by country. Successive programs in Turkey, Morocco and Yugoslavia built on earlier experiences, revising where necessary, and in the case of Yugoslavia, making a major shift from large social sector production units to emphasis on smallholder production. This shift of emphasis had also occurred in Turkey and Morocco and represented a welcome development which appeared to be the trend in all regions. The Rumanian experience represented an isolated set of conditions and as noted above was probably limited to one country".

Eastern Africa Region (EA) Dairy Projects

Bank-assisted dairy development in EA, six projects in four countries (Ethiopia, Kenya, Tanzania and Zambia) and a modest component in a Malawi rural development project, was small. Total project costs amounted to US$88.5 million, total dairy components amounted to US$65.4 and IDA credits amounted to US$56.6 million.

The Ethiopian project was greatly affected by major political changes and was generally unsuccessful although the country has a good potential for dairying. The dairy components of three Kenya smallholder credit projects were highly successful.

The extent of success is indicated by a consumption level of 75 litres/capita, by smallholders supplying 75% of the total market supply. The Tanzania project supported large units (350 cow units) under parastatal management and since 13 out of the proposed 17 units established did not cover operating costs, because costs were high and production coefficients were much lower than appraised estimates, the project was a failure. Although a smallholder component was included in the Tanzania project, it was seriously constrained by a shortage of grade cows. The Zambia project was drastically revised downwards after a review and the revised project involving 150 smallholders instead of the original 1,800 performed satisfactorily.

South Asia (AS) Dairy Projects

Total investment in dairy development has amounted to US$547 million, total project costs amounted to US$556 million and loans and credits amounted to US$250. Four dairy projects in India and one in Sri Lanka were supported. In addition two livestock projects in Burma and Pakistan had dairy components.

The Indian dairy projects have been singularly successful. Dairying in India is characterised by the development of well-managed cooperatives which handle collection, processing and marketing, and provide support services efficiently to existing small dairy farmers, who typically own one or two milking buffalo. Although production conditions in Pakistan are similar but generally superior to those in India and although the project aimed at replicating the Indian Amul model as far as possible, its performance has fallen far short because the project was not as well managed and was not as successful at institution building, at commanding government support or at defining and implementing appropriate pricing and marketing policies.

The Sri Lanka project was drastically revised after initial disappointing experience to replicate the Indian/Amul model as far as possible. After revision, the performance was fairly satisfactory. A project feature was the poor performance and high mortality of heifers imported from Australia. The Burma project had limited success because the Socialist government showed little interest in supporting and developing the smallholder dairy sector despite its considerable potential.

East Asia Pacific Region (AE) Dairy Projects

The regions total investment in dairy production was US$50.3 million, total project costs were US$122.6 million land Bank loans amounted to US$62 million.

Two dairy projects were assisted in Korea. A smallholder coconut development in Malaysia had a fairly substantial dairy component costing US$13 million and under the Philippines Second Livestock Project a small pilot dairy component costing US$0.225 million was supported. The Korean projects were highly successful in financial terms and production coefficients in most cases reached or exceeded appraisal expectations. However, they were judged to have negative economic rates of return by the Bank when opportunity costs of imported products were taken into consideration.

Under the dairy component of the Malaysian project the importation of 6,600 heifers for Government raising centres was envisaged. Serious problems were encountered with imported heifers, pasture development was much slower than expected and government tended to support large scale public and private enterprises over smallholders. A Bank supervision mission recalculated the ERR and showed that it was negative.

It was shown that milk, reconstituted from imported ingredients, cost US$0.22/litre compared with US$0.24/lither for local milk delivered to the processing plant. On the basis of this analysis, government was advised to slow dairy development and most of the available funds were not used. Malaysia has an extremely humid tropical climate and Bank staff consider that the potential for dairy development is extremely limited. The Philippines pilot dairy component had limited success. Although the Philippines has some potential for dairy development despite the humid tropical climate, Bank staff are of the opinion that dairy development there will be slow.

Review of World Bank Livestock Activities in Dry Tropical Africa

The 1981 "Review of World Bank Livestock Activities in Dry Tropical Africa" covered 34 livestock and 37 mixed livestock/crop projects in 26 Sub-Saharan countries. It was undertaken by Mr. Stephen Sandford who had considerable experience of livestock development in Africa. The review dealt primarily with 30 "livestock only" projects (in 22 different countries) and only cursory reference was made to 37 mixed projects. Although the review was based on information in Bank Documents, the author drew heavily on his independent knowledge and broad experience of livestock production and pastoralist in arid and semi-arid regions.

The report concluded that "ranching" projects or ranching components failed dismally. It also concluded that components to improve marketing and livestock movement, slaughtering and processing "have an abysmal record" and that veterinary components and off-range fattening by smallholders have a "generally good record". Sandford was unable to show any statistical relationship between failure and the size and complexity of projects, but nevertheless concluded" it is my belief that projects are too big, too complex and excessively dependent on expatriates". He further concluded " that the increasing size of projects during the 1970's was more related to the Bank's own desire to spend more on agricultural sectors than on a realistic assessment of the absorptive capacity of the livestock sub-sector". He also argued that the Bank emphasizes what happens at the top on the performance of the bureaucracy and underemphasizes what happens at the bottom actual performance at the field level.

Role of the Bank

In Sandford's view, the Bank, apart from providing capital, provides three other important elements: (a) pressure on governments not to neglect their livestock sectors; (b) specific pressure in favour of particular policies, programs and components; and (c) technical advice on particular points. Overall he felt that little of the increased meat and milk production during the previous 20 years could be attributed to Bank projects or government programs if rinderpest measures and water development were excluded.

He considered that what development there was came from a growth in livestock populations, required and made possible from growth in the human population and some expansion in livestock forage (crop residues). Despite this, he cautioned that livestock production and the welfare of livestock owners will decline further "... unless more effective and more wide-scale government sponsored programs are undertaken" because of the decreased availability of land for extensive production systems and the encroachment of cultivators on grazing land.

Despite his criticisms, he believed that the effect of the Bank's involvement on livestock development had "been beneficial" although over-influenced by "fashions" such as ranching in the '60s, fattening in the early '70's and group formation more recently. He concluded "... Bank-financed projects are usually better oriented, as well as better financed, than other livestock programs implemented by governments, and the visits of Bank appraisal and sub-sector review missions are often the occasion on which government think most deeply about their livestock policies and programs".

Although he concluded that "... Bank expenditures (US$750 million) on livestock development will not be justified by production increases in the short term, important lessons can be learned from the experience" and management cadres in African countries are being slowly build up". Livestock programs in Africa should be less capital-intensive, smaller amounts should be spent more slowly and much greater flexibility should be permitted. From the viewpoint of African welfare and economic growth he felt that it would be a pity "... if the Bank were to conclude that if it cannot spend very large sums fast, then it has no proper role in the future in African livestock development".

Rangelands

Sandford feels that there is little reliable evidence "...indicating either that widespread degradation is going on in African rangelands or that, if it is, it is due to livestock development programs". However, he felt that continued caution over the development of water supplies was warranted, but as much for social as for environmental reasons. He considered the Bank's increased willingness to finance veterinary components favourably because of the positive effects on the welfare of poor and rich stock owners and because he does not accept the environmental danger argument. He condemns "ranching" as opposed to pastoralist systems on many grounds including:

- ranches are less equitable because they increasingly favour the rich and powerful;
- they are often a reason for expropriating land which is already being fully or partially used;
- at least, under African conditions, ranches have no advantage in terms of increased food production over pastoralist systems or smallholder herds;

- at least in terms of production coefficients there is no reason to favour one type of ranch (pastoral, cooperative, company group or private) over any other commercial arrangement by pastoralist; and
- there is little evidence that animal numbers (stocking rate) can be controlled better on ranches, including private ranches than under pastoralist systems.

Sandford does not agree despite many claims to the contrary "that the know-how already exists to improve range and pasture productivity in African arid regions (less than 600 mm)". Although he appears to agree that technology exists to improve rangelands and productivity in the higher rainfall areas (over 600 mm) major questions remain to be resolved, e.g., is technology cost effective and how can the management capability needed to successfully implement it be developed? (In view of the Bank's experience with ranches, and its justifiable reluctance to support them, the question is probably moot at this stage except for smallholders).

Sandford placed considerable emphasis on the need for increased support for livestock research and attention was drawn to the smallness of research components in Bank projects (about 1.4% of total project costs). A recommendation was made "that the Bank should allocate a substantially higher proportion of its livestock programs for research" and it should adopt a more active policy of assisting research units in major livestock countries.

In view of the Bank's negative experience with livestock marketing and processing components Sandford recommended a special study on these. Much more attention should, in his view, be given to ranch records and information on ranch performance, in the Bank's records, is judged to be grossly inadequate. Emphasis was placed on the need to make greater use of "competent anthropologists" in project preparation and implementation.

Although he refers to the use of anthropologists in 30% of livestock projects, there appears to have been no measurable difference between projects whether or not anthropologists were used. Sandford considered that veterinary components of livestock projects were successful although the extent to which costs were recovered was not clear. Producers put high priority on animal health inputs and the evidence available showed that they were prepared to pay for drugs, vaccines and medicines if given the chance to do so. He recommended that considerably more support should be given to veterinary research to fill the gap in coverage between the research programs of ILCA and ILRAD, to provide guidelines on how veterinary services should be improved and how field delivery systems should be organised and developed.

Pastoral Associations

Although the recent emphasis on pastoral associations (Sandford includes group ranches in this category) was considered "a move in the right direction", he cautions against seeking some "universal model".

Since he was sceptical about the value of grazing management or controls on animal numbers, except to the extent that pastoralist may themselves handle these matters, he was against their use as vehicles for implementing government controls on grazing or stocking rates. Association's main functions, in his view, embraced land tenure, resource management, provision of services, communication of information, external relations and the building and maintenance of community cohesion and morale. He made the point that associations have land tenure and land reform implications and they should be treated like land reform projects.

General Problems

Staff-related problems, such as the inability to recruit suitably qualified staff occurred in 90% of the projects analysed (by Sandford) and political policy-related problems in about 70%. In Eastern and Southern Africa, government policies were one of the most frequently cited problems. Formal project coordination committees worked badly, especially if established at a senior level. Project management costs, although difficult to estimate, appear to have been excessively high in some cases.

The Sandford review takes issue with the project approach towards livestock development in Africa and argues for a program approach. It also argues for flexibility in design and implementation of projects in dry regions to enable management to deal with unforeseen circumstances such as droughts. Sandford believes that the Bank becomes over-involved in detail and that Bank project appraisal methodology "has more to do with Bank ritual than with the effective design of projects".

Investment in Livestock in Developing Countries

Present Status of Livestock Projects

The performance of livestock projects is still a cause for concern. On the Bank's internal rating system, mandated for all projects under supervision, the performance of 40% of livestock projects in 1989 was unsatisfactory. Only Fisheries projects have a worse performance record and even area development projects, although bad, are rated somewhat better than livestock.

This rating refers to free standing livestock projects and must be interpreted with some caution because one cannot infer from it that livestock components in broader based agriculture projects are performing less satisfactorily than other components. Since the rating system applies to the overall project we must assume (unless otherwise stated) that the performance of the livestock component is fairly represented by the overall rating. It is important to keep this in mind when one realises that most of the lending for livestock is now made under component projects.

The findings of the three studies which I referred to are still valid and describe the main problems which underlie poor project performance. Overall performance is heavily biased downwards by poor performance, especially for Africa and to a lesser extent the East Asia and Pacific Region. It is important to note that the performance of Livestock projects in all other regions was similar to Agriculture projects in general.

The bias is considerable because the volume of lending and the size of projects are relatively small in these regions. In response to past deficiencies the lending program is now paying more attention to supporting smallholders, improving technical support services (i.e., extension, research, veterinary) institution building, cost recovery and privatization. Considerable emphasis is being placed on strengthening and restructuring veterinary services, on extension and research and encouraging the formation of Pastoral Association (for veterinary and resource management) in the Sub-Saharan Africa. Dairying is receiving renewed emphasis; dairy projects are under consideration, for example, in Sudan, Kenya, Madagascar, Uganda and Zambia. Furthermore, livestock have an important role in the "Bank's Areas of Special Emphasis", namely: poverty alleviation, food security, protecting the environment, private sector development/

public sector reform and women in development. There are, for example 26 Sub-Saharan African countries with extension and 17 with research projects, although the primary focus in these projects is on agriculture and not livestock. Much more needs to be done to strengthen livestock extension and research.

In addition to new initiatives in project design more efficient economic policies are being emphasized in structural and sectoral adjustment lending. These include more realistic exchange rates, free market policies, privatization and cost recovery.

Important Factors and Examples

Profitability

It is important that long-term markets are available (and verified) to ensure profitability. Financial projections and rate of return calculations should be conducted with care. Although the World Bank places great emphasis on project analysis it is surprising how often projects fail because of inadequate profitability. Since livestock development is a long term activity, inflation and as a consequence high interest rates cause serious cash flow problems during the earlier project years. In addition to problems with prices and markets (which should be verified with care) technical and production coefficients are often over-optimistic.

Unless judgements are made by experienced operators they are inclined to reflect what is technically feasible, or what is feasible in the developed world, without due allowance been made for differences in management capabilities, technical services or the availability, cost and quality of input supplies. For example, in African ranching or dairy projects it was assumed by all donors in earlier years that parastatals could achieve production coefficients similar to those obtained on good commercial (settler) farms. With hindsight we know this was not the case — parastatal coefficients were similar to those achieved in the traditional sector and in addition parastatals suffered from political interference, overstaffing and price regulation as well as poor management (including financial management).

Profitability is also determined by the level of capital investment. Smallholders have a substantial advantage in most cases over large commercial farms. Combined with low opportunity cost labour (family) this can give smallholders a distinct advantage over large farms and parastatals. Consequently their financial viability is considerably less threatened by low production coefficients, prices or management skills.

Economic Justifications

Even if an enterprise is financially attractive this may not mean that it is economically attractive as measured by the economic rate of return (ERR). The ERR omits subsidies and taxes. Border prices (free world market prices) are used to value inputs and outputs. Consequently, a satisfactory ERR is a much better indicator of long-term sustainability than the FRR. The ERR is particularly important in countries where free markets are not permitted to operate and where the exchange rate and prices are seriously distorted.

Government Economic Policies

An overvalued currency is without doubt the most serious constraint on orderly sustainable livestock development in many countries. It is particularly serious for countries with an export surplus or the potential to develop exports. There is for example, an excellent market in the Middle East for live animals (especially sheep and goats) but Somalia, Sudan and Ethiopia have not been able to exploit this market, except to a limited extent, because exchange rates have been grossly overvalued. Conversion of earnings at the official exchange rates can be equivalent to a tax of 100% – 300% on livestock exports.

When one considers the stimulus that a 100% real increase in livestock prices would have on producers willingness to use new techniques and increase production, one begins to appreciate the magnitude of the overvalued foreign exchange problem. Governments are reluctant to move to a more realistic exchange rate because this would affect local consumer prices, especially for city consumers who wield a disproportionate political influence.

If the exchange rate is grossly overvalued the conditions are automatically created for a flourishing smuggling trade. Large numbers of animals (especially sheep and goats) are presently smuggled live to the Middle East from East African countries and large numbers of cattle are smuggled (walked) into Kenya from adjacent countries. If these conditions are not corrected, projects, which are aimed at improving livestock and meat marketing are doomed from the outset. This is usually true even if the actual marketing operation is in the hands of private traders with the public sector role confined to the provision of marketing and processing facilities on a fee for services rendered basis. Although, it can be argued that smuggling has positive economic features it is a costly and inefficient way to conduct business. Government is denied access to revenue and foreign exchange which could have been generated by official exports.

A grossly overvalued exchange rate has an additional negative effect in that the real costs of inputs that require foreign exchange are not reflected in the price paid by farmers and are therefore, not recovered even if a full cost recovery policy (local currency) is in place. This is an important issue and common difficulty with revolving funds for veterinary drugs and medicines in donor supported projects.

The adoption of a realistic exchange rate appears to be the solution to this problem. The Bank is addressing the exchange rate issue through its structural adjustment program, albeit with mixed results. Sustainable livestock development will, in many countries, depend on the success of these efforts especially if success is directly related to the availability of foreign exchange as would, for example, be the case for a veterinary project importing drugs and medicines. There will always be a problem of this nature unless a country's currency is freely convertible at realistic exchange rates.

Government price controls are a major cause of failure for projects which rely on Government or parastatal production, processing and marketing; they can also affect the private sector if rigorously implemented.

Overall effects can even transcend national boundaries. Livestock projects in East Africa were negatively affected, to a major extent, by government price control policies — even private sector ranches as well as parastatals. Livestock specialists and planners should not feel too guilty about parastatal failures because these policies had their origins in economic management theories and embraced all sectors (e.g., coal mining, car manufacture, and transport (in European countries) in those heady economic planning days.

Technology

Sustainable technology must take the realities of the country in question into consideration. Too often attempts are made to transfer technology from developed countries without realising that they are unsustainable because labour/capital ratios are completely different and support services less developed; not only technical ones but also electrical, mechanical, manufacturing, communications and transport.

Investment in large commercial state or parastatal dairy production and processing in Africa and elsewhere is a good example of inappropriate technology. Smallholder dairying requires very little capital investment and, in addition, can usually utilise low opportunity cost labour. Consumers in many countries are not prepared to pay the extra costs of heat treating, packaging and distributing milk. Systems which are based on the sale of unprocessed milk (either fresh or sour), distributed from door to door by farmers or local vendors, are much more robust from an economic standpoint.

Consequently dairy projects based on pasteurisation and sale of milk in bottles or packages can only be justified if the milk shed is located at a considerable distance (at least 50 km) from the city market. Even where pasteurisation is justified by transport distances considerable savings can be made by selling bulk milk to local vendors; as an alternative to a marketing system which may be doomed from the outset by heavy packaging and distribution costs.

The key mistake in attempting to transfer modern dairying practices to the developing world, arises from a lack of appreciation of relative costs of labour and size of incomes which by comparison with the developed world are usually 50 to 100 times smaller. The relative cost of labour must also be kept in mind when designing appropriate Artificial Insemination and other technical and input supply services.

When one considers that the average workers daily wage is only equal to about 2-3 litres of milk, it is clear that every effort must be made to utilise cheap labour and economise on fixed capital (buildings, machinery and vehicles) as well as economising on foreign exchange expenditures. These issues are discussed in more detail in a World Bank report entitled "Dairy Development in Sub-Saharan Africa" which will be published shortly in the World Bank's Technical Paper series.

The importance of grading-up, by crossbreeding, to achieve good dairy merit, must be emphasized. Crossbreeding is the cheapest and least risky approach. It avoids problems associated with acclimatisation and minimises the disease risk.

The record shows that mortality rates are unacceptably high for European breeds imported to tropical/sub-tropical regions and the number of animals imported should not exceed what is needed to establish an efficient crossbreeding program.

Processing technology, for milk and dairy products, should concentrate on simple manual system or ones with minimal mechanisation — for milk separation, butter churning and small scale village cheese manufacture. Butter and cheese manufacturing is particularly important in areas which have a pronounced seasonal milk surplus and that are located far from liquid milk markets. Milk powder manufacturing plants can rarely, if ever, be justified because of their high capital and operating costs and the availability of subsidised milk powder on world markets.

Large processing plants for meat and poultry are also difficult to justify. Small abattoirs that provide slaughtering facilities for a fee are the best solution if the number of animals involved can justify moving beyond the private butcher. Housewives would usually prefer to use their own labour to dress fowl rather than pay high processing charges. Furthermore, live chickens are much easier to store and can be killed when needed.

It is interesting to note that this system is still operating in Taiwan where per capita incomes are several times higher than in most developing countries. In my experience we should look to Asia for appropriate models for processing and marketing livestock. Furthermore, there is now ample evidence to support the proposition that responsibility for marketing live animals and livestock products should be delegated to the private sector.

If the public sector has any role it should be confined to the provision of infrastructure (markets, railway stockyards and lairage at ports) and, in addition, help to facilitate these activities by reducing red tape, improving telephone and telex services, abolishing taxes, and facilitating veterinary certifications for exports. The working capital that is needed in the livestock and meat trade is extremely large and Governments must ensure that adequate amounts are available through the banking system.

Feed Supplies

The sustainability of smallholder livestock depends, in large measure, on feed supplies. Virtually, all smallholders produce forage and crop by-products and therefore dairying and cattle/sheep fattening enterprises can be undertaken by virtually all farmers. Furthermore, even if the quantity of feed is small it usually has little if any cash value (except when sold or bartered to nomads). Since home produced feed costs little, if anything, smallholder livestock farming is unburdened of one major risk-that associated with large fluctuations in feed prices and reliance on purchased feeds.

Likewise smallholder pig and poultry systems flourish on farms producing grains (e.g., maize, rice) and crops and cereal by-products (e.g., rice bran). Robust smallholder pig production was a feature of Danish agriculture until very recently (a flourishing pig production industry based on home grown barley) and robust smallholder pig production still flourishes in Poland (based on home grown potatoes and cereals) and in Yugoslavia (based on home grown maize and wheat pollard from home grown wheat).

Likewise smallholder pig and poultry production flourishes in Asia because all smallholders have access to cheap rice bran. Rice milling is usually dispersed through numerous small villages and farmers receive or buy back most of the rice bran from their own crop; rice bran stores badly because its oil content is high and it is therefore unattractive to feed compounders. It must be fed fresh and since rice milling is a continuous process rice bran is constantly available to smallholders.

These traditional smallholder livestock systems are virtually risk free and will survive (as they did in Western Europe) until wage levels are such that their contribution to family income becomes insignificant or when they are incapable of providing an economic labour wage. For example, in Asian rice systems, a man can manage a flock of about 150 ducks grazing on rice paddies.

This system persists when the cost of feed saved, by scavenging, is sufficient to justify a man's wage but the practice has collapsed in countries where wages have surpassed this level. These examples are given in order to stress the point that one needs to be vigilant and fully understand the implications of the underlying economic realities as well as their influence on switching points in farming systems (e.g., from manual to machine milking).

Smallholders are economically more robust than large commercial producers because they are much less subject to feed and product prices. They should be encouraged to the fullest whenever possible and will persist until gradually surpassed by economic development. Feed is normally the binding constraint on smallholder production and projects designed to assist this sector should always incorporate a well thought-out feed and/or forage component.

Commercial sustainable pig and poultry industries can be developed on imported feeds in countries with a deficit in these products (e.g., Taiwan, the Philippines and Korea). This is possible because it is much cheaper to transport feed grains than meat and poultry products. However, if the industry is to survive in the deficit country production standards and efficiency must be comparable with those found in developed countries. Developing countries usually has a substantial advantage in labour and construction costs and, in addition, the fertilizer value of waste products is usually much higher and as a consequence the waste disposal problem is minimised.

Although somewhat surprising, it is worth noting that few countries have a well thought out strategy for meeting their short, medium and long term feed supplies (energy and protein feeds). One needs only to look to the continuing chronic feed protein deficit in Eastern European and Asian countries to realise the full magnitude of the problem. Perhaps FAO and the World Bank could play a more substantial role in rectifying this situation. The gains that can be achieved, if enlightened policies are put in place, are phenomenal and clearly evident from the expansion of pig and poultry production, on imported feeds in many developed and developing countries.

Sustainable Support Services

Sustainable smallholder development depends on sustainable technical and other services. Sustainability of services is largely a function of cost recovery. Even where the principles of cost recovery and privatization are accepted the task of developing sustainable institutions to deliver these services is a formidable one. Considerable investment will be needed in institution building, technical assistance, management training and staff training at all levels. Grass-root farmer organisations (i.e., pastoral association and cooperatives) need in most cases to be established and farmers trained to own, operate and manage them. This is a task that goes much beyond the life-span of the typical livestock project. One should look to Operation Flood in India and village agriculture/livestock cooperatives in Taiwan to see what can be achieved in the mature stage, and to initiatives for restructuring veterinary services and establishing pastoral associations in Sub-Saharan Africa to see what can be achieved at the earlier development stages.

In recent years the Bank has placed much more emphasis on institution building and the provision of services (e.g., extension, research and veterinary), in full cognizance of the importance of these to sustainable development. Although this is not the place to discus these important services, it is pertinent to point out, that in most developing countries livestock services are either non-existent, weak, non-effective or absent. The design of efficient affordable services is a difficult task which calls for innovation, the rejection of old nostrums and careful cost/benefit analysis to establish affordability, a pre-requisite for sustainability. While donor agencies can assist this process, governments must ensure that coherent policies and strategies are implemented to avoid confusion and to save time and resources. Donors must ensure that their activities and projects are consistent with the policies and strategies which have been set by Government.

Animal Health Services are of paramount importance to the sustainability of smallholder system but especially to dairy farmers using disease susceptible crossbreds. Although good progress has been made on conceptualising restructuring and cost recovery most livestock farmers in Sub-Saharan Africa have still to 'make do' with, at best, a rudimentary and, at worst, a totally ineffective service. In Ethiopia, for example, the expenditure on veterinary drugs and medicines is only about 5% of the amount veterinarians estimate would be economically justified on the basis of epidemiological studies. It will be extremely difficult to rectify this situation even though the principles of restructuring and privatization are fully accepted. A major effort is needed to train veterinary field assistants that will be employed by service cooperatives, the front line institutions. Systems of bookkeeping and cost recovery must be developed and demonstrated which require time and a massive training program. A mechanism is needed to enable payments in local currency to be converted to foreign exchange to replenish imported veterinary stocks. In addition procurement and distribution must be streamlined.

Although one must assume that these problems can eventually be resolved one should question if the contribution or the role of farmers in administering drugs and medicines is adequately taken into account in present Sub-Saharan Africa models where veterinary assistants, selected from and paid by the traditional village community or pastoral association, are responsible for administering drugs, medicines and vaccines to livestock. When one considers that probably 90% of the veterinary drugs and medicines are administered by farmers in developed countries (as represented by farmer's expenditure), albeit in most cases under the veterinarian's instructions one begins to realise the importance of training the African farmer to play a much greater role in the administration of veterinary products. It would take an enormous increase in manpower, travel time and cost to replace the farmer's legitimate function in, for example, 'drenching' for stomach worms and liver fluke. Study and analysis is needed to sharpen the focus on these matters - perhaps FAO could help by establishing badly needed guidelines.

Environmental Sustainability

Despite some common misconception livestock projects normally make a substantial beneficial contribution to sustainable agriculture. In arid areas there is now ample evidence that irreversible degradation is not taking place on a large scale as a consequence of over-grazing. Rangelands generally recover when droughts give way to a wetter cycle of annual precipitation. The public clamour and fear of irreversible degradation appears to follow a similar cyclic pattern. The real environmental problem in arid areas is caused by erosion brought about by increased and continuous cropping as well as bush cutting (for fuel) which is, in turn, a consequence of population pressure. Overgrazing does exist and can cause erosion on slopes, but these effects are minor compared to those caused by human population pressure. Furthermore, grasses, forage, legumes and legume trees, introduced to provide livestock feed, are important builders of soil fertility and, in addition, provide ground cover which prevents or lessens wind and water erosion in susceptible areas (e.g., Ethiopian highlands).

Regional Policies on Food and Agriculture and their Impact on the Advancement of Livestock Production in Developing Countries an Example from Africa

Looking back on the European Development Funds (EDF) 30 years experience, particularly in Africa, in promoting increased livestock production, a number of conclusions can be drawn:

- that despite massive investment by the European Development Fund (EDF), and other donors, we are far from achieving what we set out originally to achieve. Developing countries, particularly in Africa which are the main recipients of the European Community's development aid, are more dependent on imports than ever before,

- that this is due more to the implementation of the inappropriate policies than to a lack of funding, and
- that the technical know-how that exists is not applied to the extent it should be.

In the Community we speak increasingly about policy dialogue. The newly negotiated Lome Convention (Lome IV), gives special attention to the development of viable policies to increase food self-sufficiency in the countries who are signatories to the Lome Convention.

With regard to animal production, the Community organised in 1984 a conference at which experts from the EC Member States and ACP (African, Caribbean and Pacific) countries were invited to discuss the results of the cooperation in the development of animal production. The result was a document entitled "Basic Principles for Animal Production Development". Although this document was established more than six years ago by representatives of some 66 developing countries and 12 European Community Member States, a number of these principles remain topical today.

An important recommendation was that sectoral policy concerning the development of livestock resources, should be focused on the livestock producer and his family. Individual producers should be considered as the main EC beneficiaries. By implication this means that financing large state farms or large state organisations should be avoided, instead, attention should be given to livestock associations, cooperatives and other forms of voluntary organisation. The latter point is repeated in the recently negotiated Lome IV Convention.

Other basic principles of Lome IV include:

- That every effort be made to promote and encourage private initiative in the provision of goods and services needed by the livestock producer.
- That it is essential to formulate and implement a pricing policy for both inputs and livestock products alike. The cost of inputs must eventually be paid in full by the user, although this does not rule out the possibility of a subsidies policy of limited duration to encourage the use of certain inputs.

The necessity of developing appropriate livestock policies was seen at the beginning of the 1980s when rinderpest broke out, once again, on a large scale in Africa.

Despite all the effort and investment that went into JP 15 and earlier rinderpest campaigns and when it was thought that rinderpest had almost been eradicated; it was a surprise to many how easily and quickly the disease spread between 1980 to 1984. It appeared that the operational budgets of the veterinary services had in no way kept pace with the increase in personnel caused by the automatic recruitment of graduates. Often up to 90% of the total budget was used for the payment of salaries and what remained was insufficient to provide an efficient service to farmers.

The approach that the Community developed was twofold. In countries where actual outbreaks of rinderpest occurred, immediate aid was provided without conditions. Elsewhere the following approach was developed. First, a thorough analysis was made of each country covering: the government veterinary budget, the number of personnel, the contribution of livestock to the national economy and the self-sufficiency rate in livestock products.

The objective was to achieve a situation where sufficient finances are available either on a government budget, or on another basis, to support the programme. It was, for the Community, unacceptable that, as with JP 15, the major part of the funds would come from donors to be used exclusively for the financing of vaccination campaigns, in the form of vehicles, equipment, vaccines, running costs.

Despite initial opposition, the Community was lucky in finding an ally in the Inter-African Bureau of Animal Resources of the Organisation of African Unity (OAU) who shared the same philosophy. It was necessary to find a system which would increase the employment opportunities for the increasing number of veterinarians and zootechnicians being trained in Africa to find gainful employment. Furthermore these employment opportunities should not be created by the government sector alone.

In 1986 a financing agreement was signed with the OAU which, apart from those countries receiving direct support, stipulated that finance would only be available after a successful outcome of a policy dialogue. Five possibilities were identified on how to improve financing of the livestock services, these options were:

- Is it feasible for farmers to pay the full cost for all services rendered? Almost everywhere in Africa it is now accepted that farmers must pay a realistic price for the goods and services they receive. There was certainly not a universal acceptance of this practice in 1986.

The Community were aware that policies cannot be changed overnight, therefore, EDF funds are used to make a gradual adaptation of new policies possible. The question has often arisen whether or not compulsory vaccination should be free of charge. The Community taken a neutral attitude, however, if a country wishes to provide these campaigns free of charge, it insists that prices of other products services are increased to such an extent that the "free services" are paid for indirectly.

- An alternative to direct payment for services rendered, could it be to organise farmers to pay an annual contribution in return for certain services. Although similar to a medical insurance scheme, in practice, this option has not proved to be popular and only in a few countries is it applied.

- Instead of government recruiting all the qualified personnel can it be encouraged, with EDF funds, that cooperatives, farmers organisations, etc, recruit directly their own staff which will be utilised directly to serve the farmers needs. This has proved to be an attractive option.

- Similarly, the EDF funds can be utilised to establish veterinarians and zootechnicians in their own practice. Of all the options this one has proved the most popular by far and is an indication of how attitudes are changing in Africa.

In practice, a bank in a particular country which is best suited to manage the loans is selected. Subsequently, EDF funds are used to a guarantee loans for aspiring candidates to set up in practice. In certain specific cases EDF may discuss the possibility of loans at a reduced rate of interest.

- The last and the most controversial option, is to tax the livestock sector with part of the revenue put into a special fund called a Livestock Development Fund. Also, if livestock products are imported at a price which is not possible for efficient local producers to compete with, then levies should be imposed and again these levies should be deposited in the Livestock Development Fund. A number of countries proved to be quite receptive to this idea but it appeared that IMF conditions for structural adjustment often prevent this. Ideally, the result of structural adjustment will be sufficient local budgetary resources available for livestock services and the need for a Livestock Development Fund would no longer exist.

Finally, you will have noted that I have emphasized the words "livestock services". The whole object of the Pan African Rinderpest Campaign is, ultimately, to increase animal production to ensure greater food security. This cannot be done by putting the emphasis on animal health alone, we must give equal attention to the problems connected with improved husbandry and especially animal nutrition. The financing agreement must, therefore, mention specifically likely consequences on the environment, particularly decertification, and, where necessary, ensure special action is taken to correct the situation.

The above is an example of a European Community approach to livestock development in Africa and, subsequently, with similar projects in India. With regard to specific production goals less clear cut policies have been established. The approach taken is that each situation has to be considered in context. But the general guidelines are as follows.

- That more attention must be paid to milk production in developing countries. The schemes undertaken by the EC in the past have not achieved sufficient results. In developing dairy production it is necessary to start with the local animal and that their production potential is exploited to the full. In general, we can assume that the genetic potential is sufficient, however, in order to cope with local demand it will be necessary to introduce sooner or later a degree of "foreign blood". Crosses with European dairy breeds are usually recommended. It is envisaged that privatisation schemes for veterinarians mentioned earlier could also provide a private artificial insemination service as well. In the development of milk production, attention must be given to a) animal nutrition and b) to the collection and ultimate commercialisation of milk.

- Regarding beef production, contrary to milk production, the genetic potential of the local breeds is considered sufficient. Conversion rates are at least as good as those from European breeds under local environments.

- For pork and poultry production, apart from the necessity of assuring adequate veterinary and extension services the major question is the availability of adequate energy and protein resources.

- Small stock have received less attention from both governments and donors than they deserve. The few EC projects financed in this sector have, in general, given good results. More than with other animal species the chances for production improvement are considerable.

To conclude, Lome IV like its predecessor devotes much attention to environmental issues which must be considered at every stage. It is for this reason that the Community has in recent years increasingly financed wildlife utilisation projects. There is a good case to be made for the necessity of conserving genetic animal resources also that the commercial exploitation of game animals can be environmentally and economically beneficial.

The availability of adequate funds is of course necessary for the realisation of these policies, but increasingly the EC is of the opinion that when the development policies are wrong, no amount of funds can remedy this failure.

Chapter 4

The Impact of Livestock Development on Environmental Change

Environmental change, over the next 50–100 years, due to the warming effect of the accumulation of gases in the atmosphere will clearly influence man and his well-being and will introduce change to many areas of the world.

Although there has been modelling of future climate change, the major observation is that these models are as yet producing limited useful information. Computers have the ability to consider a multitude of variables but the computer is limited by the knowledge base and magnitude of the unknowns.

In fact, it can be quickly ascertained that modelling, at best, is defining what is not known in terms of what will effect change in temperature, rainfall and sea levels, let alone sea currents, wind, sunshine hours, soil moisture and the incidence of pests and diseases of man, animals and plants.

Some prediction (no matter how uncertain) are welcomed by the population at large (e.g. increased environmental temperatures in Europe) but the uncertainties indicate that for any benefits there will also be major disadvantages, both economic and social.

For example, high temperature and increased evaporation rates and the lower resultant soil moisture content, may result in the death of large areas of planted and natural forest in Northern Europe; rain in previously dry areas may lead to large scale soil erosion and wide scale flooding of major agricultural lands in the fertile delta country will occur if sea levels increase by 0.3 metres by 2050.

The major conclusion is that disadvantages are likely to outweigh any advantages and the unknowns make it essential to put into action technologies to slow green-house gas emissions and stabilise atmospheric gases as soon as possible.

The long lag time between gas production, mixing in the atmosphere and therefore warming together with the reluctance of governments to put into practice legislation to limit, in particular, carbon dioxide production from fossil fuels, suggests that there will be a rise in world environmental temperatures of 0.5 to 1°C in the next 25–50 years. A major point is that no one country can point to a low level of "environmentally dangerous gases" production as being a reason for non-compliance with a general lowering of emissions. Virtually all countries contribute a small proportion of the greenhouse gas and it is, therefore, necessary for all countries to take action. In other words it needs action world wide and "every little bit helps".

A distinction must be made between temperature rise due to gas accumulation in the atmosphere and depletion of the ozone layer through reaction with atmosphere contaminants. The two are linked but the depletion of the ozone layer (this layer protects the animals/plants from deleterious dose rates of UV irradiation) is largely a result of reaction of the ozone with atmosphere contaminants.

Ozone depletion is not highly related to the subject of environmental change and gas accumulation in the atmosphere as discussed here. However, increased ultra-violet radiation at the earths surface will have major detrimental effects on plant growth.

The Greenhouse Effect

A Simple Description

The greenhouse effect, or increasing world temperature, is clearly ascribable to the major industrial countries as some 50% of the increased retention of energy by the atmosphere is a result of the accumulation of carbon dioxide from combustion of fossil fuel.

Industrialised countries presently use 70% of the world's oil production and it has been much higher in the past.

The other gases that contribute to increasing temperatures arise from a variety of activities and some of the gases have been created by man. Methane is an important component of the increasing gases in the atmosphere and is the one most associated with animal agriculture. Methane has a thermogenic effect some 4–6 times that of carbon dioxide.

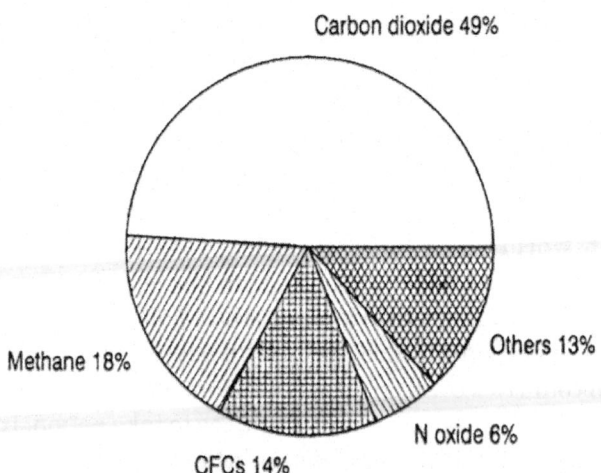

Figure 1: Relative contribution (%) of greenhouse gases to atmospheric warming (Source: World Resources Institute).

The rates of accumulation of methane and carbon dioxide in the world's atmosphere have changed dramatically in the last 10 years. Prior to this, the rise in world temperatures and composition of the atmosphere had changed little, but is now in what appears to be an exponential period. Undoubtedly the contamination of the atmosphere with carbon dioxide, methane and other gases must be stabilised or the future of the earth is threatened.

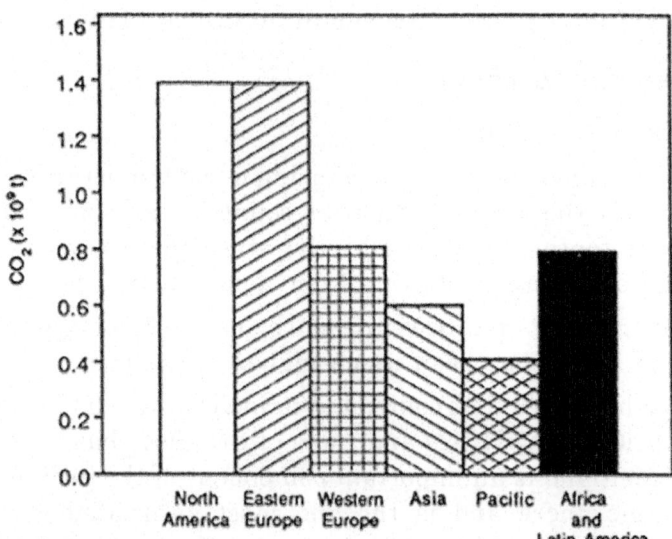

Figure 2: Relative contribution by continent to the emissions of carbon dioxide (Source: World Resources Institute).

Methane production appears to be a major issue although it presently contributes only 18% of the overall warming. It is accumulating at a fast rate, and is apparently responsible for a small proportion of the depletion of the protective ozone layer. Methane arises largely from natural anaerobic ecosystems, rice paddies and fermentative digestion in ruminant animals.

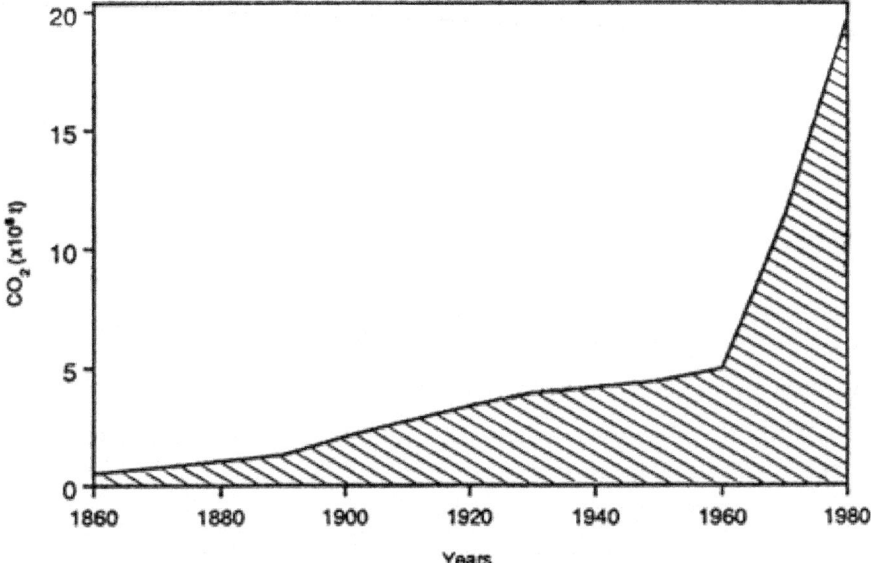

Figure 3: Trends in emissions of CO_2 (Source: World Resources Institute).

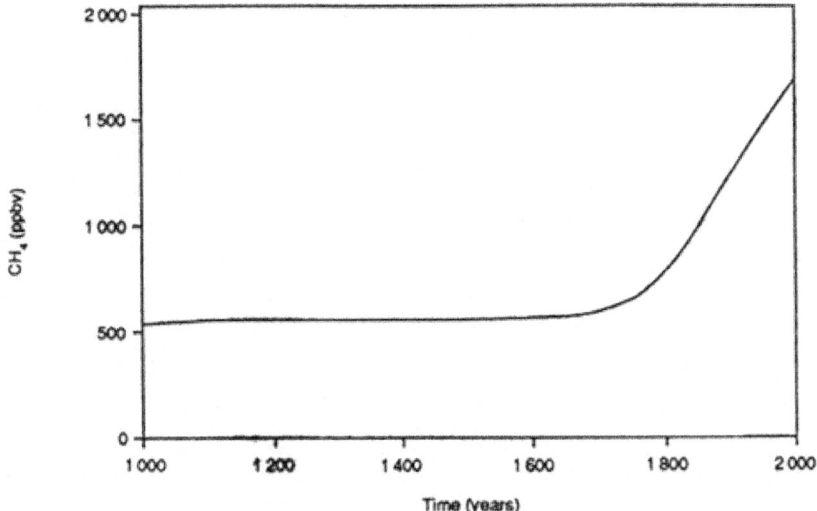

Figure 4: Trends in atmospheric methane accumulation (Khalil and Rasmussen, 1986).

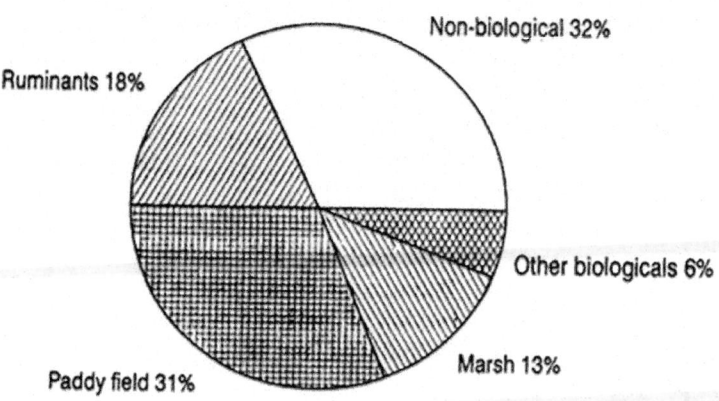

Tg CH$_4$ per year

Figure 5: Relative contribution of biological resources to the global production of CH$_4$ in the atmosphere (Bolle *et al.*, 1986).

Animal Agriculture and the Greenhouse Effect

Animal production plays four important roles in the release of gases into the atmosphere:

- directly through production of methane in fermentative digestion of ruminants,
- indirectly when a proportion of the faecal materials decompose anaerobically,
- indirectly through CO$_2$ production from fossil fuels to provide the production and marketing infrastructure and inputs such as motorised transport, fertilizers, herbicides and insecticides,
- through the clearing of forests and range lands, the timber on which was a natural sink for carbon dioxide.

Ruminants in 'natural' production systems are inefficient and, in general, production increases depend on an expansion of numbers. There is a growing appreciation that efficiency per animal can be improved many fold with simple technology inputs which would have an impact on all four aspects of the contributions to global warming discussed above.

The most important approach to be discussed in relation to the amelioration of greenhouse gas production by ruminants is to increase the efficiency of animal production from available resources and develop the capacity to produce more from less animals.

Requirements For Animal Products

Developed Countries

With largely stabilised human populations in the industrialised countries and a high standard of living generally, the demand for animal products has plateaued or declined and the emphasis in animal agriculture is on the production of higher quality products. This has led to a dependence of ruminant systems based on concentrate feeds, a higher production rate per animal and reduced animal numbers. For example, there has been a marked increase in milk production through "better nutrition" relying on concentrate diets and genetic improvement of dairy cows. The result has been a marked reduction in the total number of animals required to supply local milk requirements. These increases in production per animal in North America and EEC countries that have arisen from simple technology inputs are indicated.

The demand for food in the industrialised world is stable but production is increasing at 1.5% per annum (FAO, 1986). Policies to take land out of farming or to limit farm management options and animal numbers has reduced overall animal numbers, but meat/milk production has been maintained by technologies that increase the efficiency of production. From a global warming view point this is mostly advantageous, although there is a very high environmental cost of feeding high grain based diets to ruminants. Such concentrates can only be produced by high fossil fuel inputs and is often only economic because of subsidies. Les intensive production systems depending on grass or grass products in developed countries exhibit some of the same problems of low efficiency as those animals in developing countries which are restricted to roughage based diets.

The Developing Countries

Throughout the last 30 years crop and livestock production has more than doubled throughout the third world, although there are large differences between regions, however, increases in human population, Urbanisation and improved income levels have increased demand for food to such an extent that surpluses are still rare. Imports of meat, milk and cereal grains in most developing countries are increasing by about 10% per annum. The demand for animal products will continue to increase for some time, if the patterns of food consumption by people in developing countries follows the patterns that occurred in developed countries.

Increases in animal production in the developing countries has been mainly a result of increasing animal numbers (Jackson, 1981). The lack of increase in efficiency of animal production is well documented and is emphasised by the average milk yields per cow over the 10 years from 1976–1986 (Brumby, 1989). This low productivity is exacerbated by long calving intervals and a late age at puberty in the cows in developing countries. It should be emphasised however, that it is also a feature of ruminants fed low quality forages in any country.

New feeding strategies for animals fed on low quality forages (e.g. crop residues, tropical pastures etc.) coupled with better genotypes, improved management and disease control, particularly in India (NDDB, 1989), has changed this situation enormously.

Table 1: The change in the average milk yield per cow in industrialised and third world countries.

Country/Region	Average yield/ cow(kg/year)		Percentage increase (%)
	1976	*1986*	*(1976 to 1986)*
North America	3,250	5,200	60
EEC Countries	2,900	4,100	35
Asia	620	700	13
Africa	322	354	7

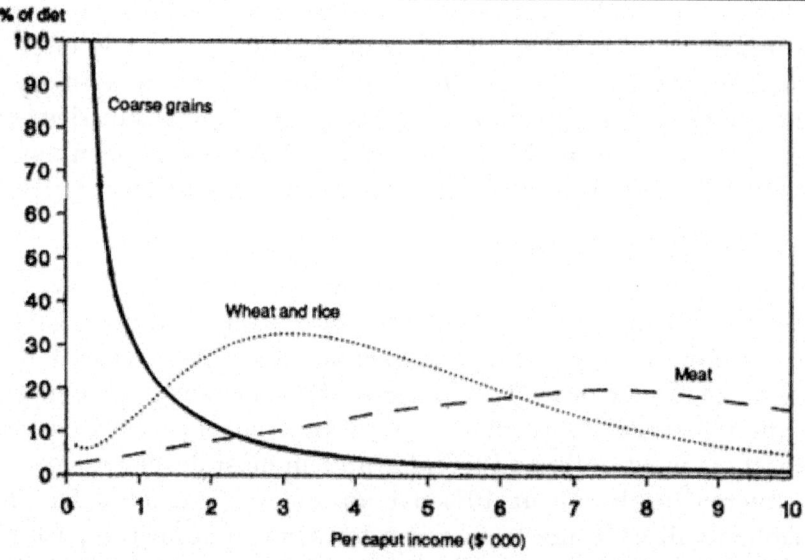

Figure 6: Food consumption/percentage of diet for meat, wheat & rice and coarse grains. (Marks & Yetley, 1987 and Brumby, 1990).

A major difference in approach to feeding ruminants in tropical developing countries that has been recently developed has the potential to revolutionise ruminant production from forages. For example enormous increases in milk production can be achieved in the tropics without the use of 'fossil-fuel-expensive' grain based concentrates, relying rather on byproducts of agriculture.

These are the only truly available feed resource for the large ruminant populations in the foreseeable future. The low inputs of concentrates into such systems and levels of production that rival those achieved in the industrialised world, provide major indications for change in management of both developed and developing countries alike.

Population Densities of Large Ruminants

The above discussions suggest that ruminant populations are not likely to increase in the industrialised world and, with ever increasing technology inputs, numbers are set to decline (as has already occurred with milking herds) - but continuing technology inputs will be highly fossil fuel dependent.

Conversely, there is likely to be a huge increase in demand for food products in the countries that are developing, of which a greater proportion of the demand is likely to be for animal products. To meet this demand local production must be increased to an extraordinary extent and it is most desirable that the impact on the environmental contamination is minimised.

The ranges of Y_{ATP} are shown for: The demand for draught power in countries with large numbers of small farmers is likely to expand rather than contract in the future and must not be left out of any considerations. This large group (80 million draught oxen in India alone) receives the least inputs and yet they are probably one of the major factors in food production and they limit the use of costly fossil fuels in developing countries.

Environment-friendly development of livestock production systems demand that the increased production must be met by increased efficiency and production per ruminant and not through increased numbers.

The need to increase numbers would put huge pressures on many resources including forests and land that might be afforested.

Methane Production

Methane Production from Ruminants

World ruminant population densities and estimated methane production rates are shown in table in comparison to some monogastric species including man.

Global Methane Production

Methane gas accumulation is at a rate of 1% per annum and methane contributes 18% to global warming.

Ruminants from all species produce a relatively small proportion of the global production i.e. 15–20% of the total methane generated. However, the domestic ruminants represent one of the few sources that could be manipulated. It is estimated that beef and draught animals contribute 50%, dairy cows 19% and only 9% is from sheep (Crutzen *et al.*, 1986). The methane is generated largely in the fermentative digestion of feed by microbes in the rumen.

The data in table indicate an approximate even split of numbers of ruminant animals between the developing and developed countries; that large ruminants contribute the greatest to world methane accumulation, and that other domestic and wild herbivores and man contribute an insignificant proportion.

The simple approach to the calculations adopted by Crutzen *et al.* (1986) may, however, be a little misleading although they are the best estimates available.

It is most important to emphasise that it is the rates of methane production per unit of product produced over a life time which is important in identifying where it is possible to make a major reductions in methane emissions. This is the major philosophy developed in the rest of this presentation.

Productivity of Ruminants Fed Poor Quality Forages

The vast majority of ruminants in developing countries and a major proportion of the national herds of industrialised countries are supported on the by-products of agriculture or graze forages of relatively poor nutritional value.

In general, growth rates, milk production and reproductive rates in these systems depending on forage of variable quality are extremely low compared with the genetic potential of these animals (mostly about 10% and rarely exceeding 30%). Mostly cattle grow to maturity or slaughter weight over 4–5 years, cows produce their first calf at 4–5 years and then, on average, every two years.

Milk production on these feeding systems is often below 1000 litres/lactation. Cows may be kept largely to produce draught oxen and in some specialised systems they are kept for the production of dung (which is valued as a fuel) and a number of other minor purposes (e.g. as a investment, for recreation and for religious purposes).

Table 2: Estimates of methane emissions from animals (adapted from Crutzen *et al.*, 1986).

Animal Type and Region (X10⁶)	World Pop.	CH₄Prod. (kg/hd/yr)	Total CH₄Prod.
Cattle:			
Developed countries	573	55	31.8
Developing countries*	653	35	22.8
Buffaloes	142	50	6.2
Sheep:			
Developed countries	400	8	3.2
Developing + Australia	738	5	2.4
Goats	476	5	2.4
Camels	17	58	1.0
Pigs:			
Developed countries	329	1.5	0.5
Developing countries	445	1.0	0.4
Horses	64	18	1.2
Mules & Asses	54	10	0.5
Humans	4670	0.05	0.3
Wild ruminants and large ruminants	100–500	1–50	2–6
Total			76–80

* includes Brazil and Argentina

** total estimate for emissions from domestic animals has an uncertainty factor of ± 15%

Slow growth, low milk yield and poor reproductive performance results in poor feed conversion and a large methane output relative to product output.

Methane Production in Ruminants Fed 'Poor Quality' Forages

Methane output relative to product output of ruminants depends on two factors:

- the efficiency of fermentative degestion in the rumen, and
- the efficiency of conversion of feed to product (e.g. milk, beef, draught power).

Efficiency of Rumen Fermentation

Digestion of feed by ruminants depends on a diverse group of microorganisms in the rumen. These organisms ferment feed materials into volatile fatty acids (VFA) a process that produces methane (CH_4), carbon dioxide (CO_2) and utilises the energy (ATP) derived to convert feed to microbial cells. The partitioning of the feed components into VFA or microbial cells and the release of CH_4 and CO_2 depends on a number of factors. Feeds that allow a high efficiency of microbial cell synthesis produce low amounts of methane per unit of feed digested.

With cattle fed on a poor quality forage a number of essential microbial nutrients are usually deficient in the diet and microbial growth efficiency in the rumen is low. In these conditions CH_4 produced may represent 15–18% of the digestible energy and correction of these deficiencies may reduce this to as low as 7%. The relationships between products of fermentative digestion and the efficiency of the microbial ecosystem in the rumen.

Efficiency of Feed Utilisation by Ruminants Fed Crop Residues or Other Fibrous Feeds

Liveweight gain. Research in the past 20 years has clearly illustrated that supplementation of cattle on low quality forage based diets effects productivity through increasing efficiency of feed utilisation. A mixture of nutrients as can be supplied for instance in a molasses urea multi-nutrient block/lick ensures an efficient microbial digestion in the rumen. Also small amounts of protein meal that are directly available to the animal (i.e. bypass protein) stimulate both productivity and efficiency of feed utilisation.

Traditional feeding standards are based on the metabolisable energy (ME) content of a feed. The general relationship between ME/kg of feed and growth (g gain/unit of ME intake) are shown. The results of a number of feeding trials with cattle on straw or low quality pasture and silage based diets supplemented with protein meals are shown in the same. The efficiency of growth and methane production is shown. These data clearly show the massive reduction in methane production per unit of liveweight gain that is possible by using a strategic supplementary feeding system that accommodates the requirements of the rumen organisms and balances the absorbed nutrients to the animals requirements. The data show the effects on methane production of balancing the rumen and for protein supplementation calculated for experimental data with growing animals of Saadullah (1984). This indicates the massive potential reduction in methane production per unit of liveweight gain that can result from supplementation (Leng, 1989). Provision of molasses urea blocks to draught oxen would have a major effect on methane production, reducing it to half the present production rate.

Milk Production

The same principles of supplementary feeding have been found to stimulate milk production of dairy animals. Without going into detail, the methane produced per unit of milk produced under traditional feeding of local dairy cows or imported Friesians in India or Friesians under temperate country management production in relation to lifetime milk production and includes the effects of supplementation on age at first calving, intercalving interval and improved milk yield of supplemented animals.

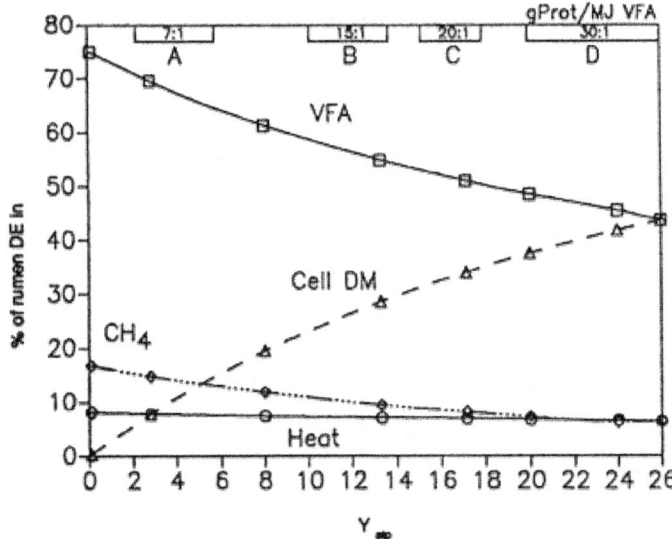

Figure 7: Schematic relationship between diet quality (metabolisable energy/kg dry matter) and food conversion efficiency (g liveweight gain/MJ ME) (source Webster, 1989).

The relationships found in practice with cattle fed on straw or ammoniated straw with increasing level of supplementation. Australia (Perdok *et al.*, 1988), Thailand (") (Wanapat *et al.*, 1986) and Bangladesh (¡%) (Saadullah, 1984).

Recent relationships developed for cattle fed silages supplemented with fish proteins (Olafsson and Gudmundsson, 1990) (Å) and tropical pastures supplemented with cottonseed meal (Godoy and Chicco, 1990) (*) are also shown. This illustrates the marked differences that result when supplements high in protein are given to cattle on diets of low ME/kg DM.

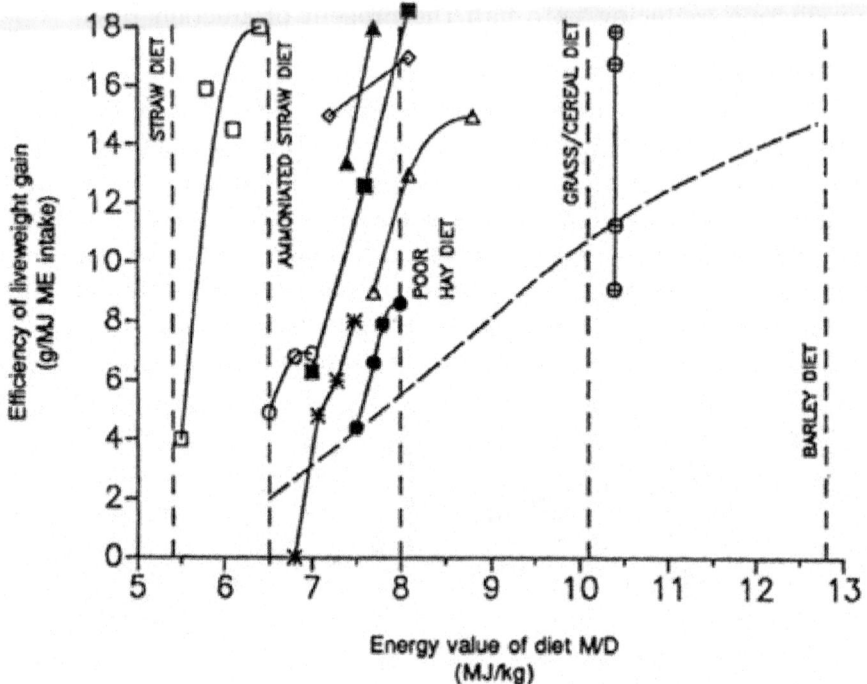

Figure 8: Schematic relationship between diet quality (metabolisable energy/kg dry matter) and food conversion efficiency (g liveweight gain/MJ ME) (source Webster, 1989).

The relationship shown by a broken line is based on the metabolisable energy system in practice in the U.K. (Webster, 1989). The other relationships are results from results quoted. Perdok *et al.*, 1988, Saadullah, 1984, Wanapat *et al.*, 1986, Godoy and Chicco, 1990 (²%) and Olafsson and Gudmundsson, 1990 (f&).

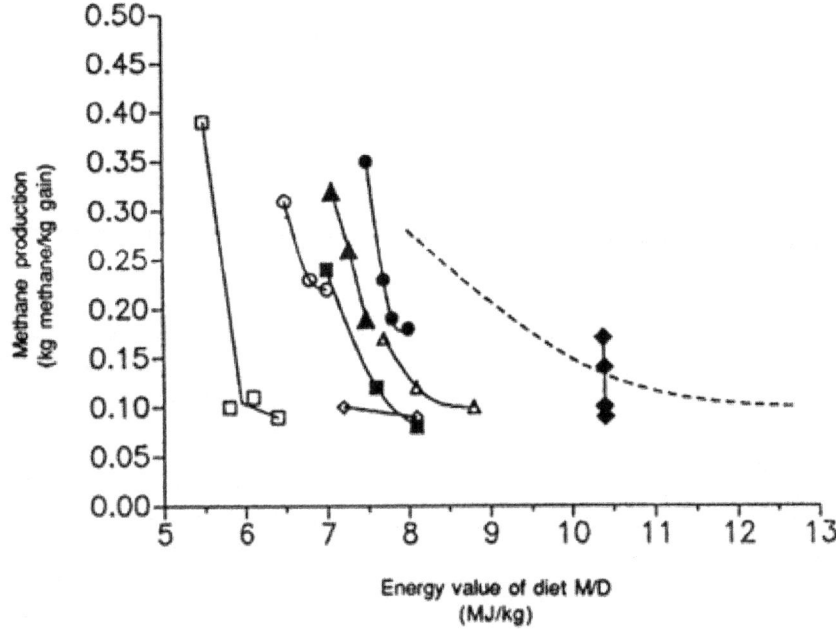

Figure 9: The relationship between the metabolisable energy content of a feed (M/D, MJ/kg) and the methane produced/kg gain.

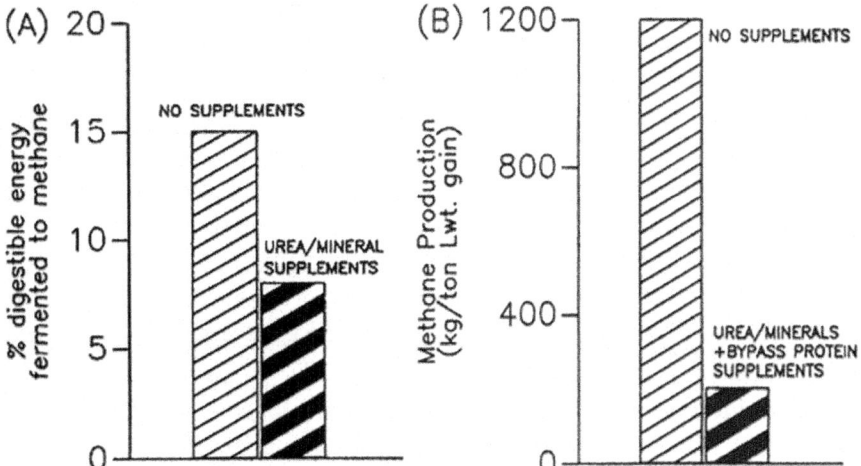

Figure 10: (A) The effects of improving the efficiency of rumen fermentative activity on methane production/kg of digestible energy consumed. (B) The production of methane/kg gain in supplemented cattle (feed conversion efficiency (FCR) 9:1) or unsupplemented cattle (FCR=40:1) fed straw based diets (after Saadullah, 1984).

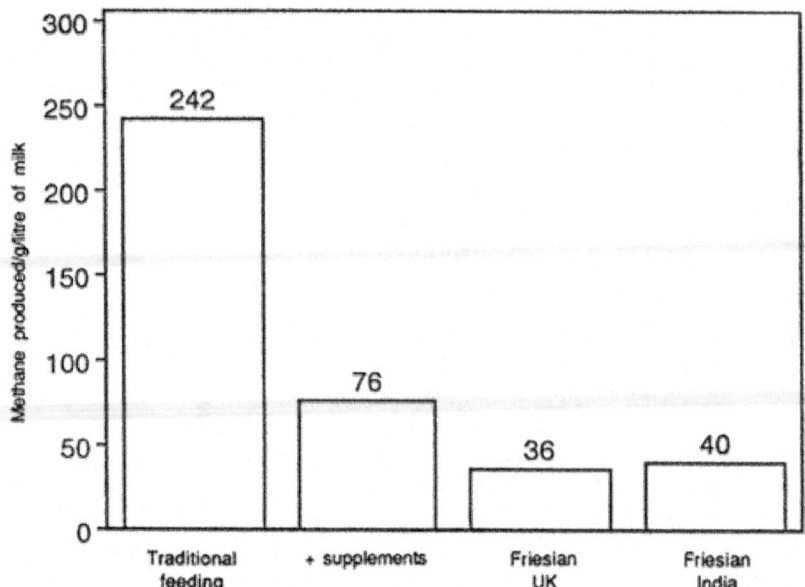

Figure 11: The methane produced per unit of milk production in unsupplemented (fed traditionally) or supplemented (new feeding systems) cows in India with moderate levels of production.

Methane gas content of the atmosphere is increasing at 1% per annum. To stabilise methane in the atmosphere, global methane production needs to be reduced by 10–20%.

Large ruminants produce some 15–20% of the global production of methane. Ruminants on low quality feeds possibly produce over 75% of the methane from the world's population of ruminants. Supplementation to improve digestive efficiency in these animals could at times halve this methane production per unit of feed consumed. Together with supplementation to improve efficiency of feed utilisation and increase product output may thus reduce methane production per unit of milk or meat by a factor of 4–6.

Provided animal numbers in national herds are decreased as demand is met, the production of methane from the large populations of animals fed poor quality forages could be reduced to below 50% and perhaps even to as low as 20% of its present rate.

It is probably not feasible to restructure the cattle industries of the countries involved. The socioeconomic implications are difficult to predict, but the start made in India with up to 1 million head of cattle, and with a potential to reach 7–10 million animals in the immediate future, is highly desirable because it increases milk production with a concomitant decrease in actual methane production and methane production per litre of milk.

Social, anthropological, economic and political considerations have been the major determinants of 'aid programmes' but the growing 'environmental crisis' due to increasing content of gases in the atmosphere leading to global warming is likely to dominate many issues confronting aid-agencies in the future.

The achievements in India can be repeated in most countries that depend on poor quality forages and even silages for ruminant production. India is fortunate in having large amounts of protein meals as by-products which can be used for animal production. In other countries protein meals may be scarce or under-utilised. The future for increased animal production in these countries is to identify protein sources and proceed to find economic mechanisms for producing meals that contain a high proportion of bypass protein. The developed world may need to encourage this by subsidising such activities or by refraining from draining _these protein sources from the developing world to support subsidised over-production in the industrialised countries. The world trade in protein meals is shown in table.

Table 3: The trade in oilseed cakes between industrialised and third world countries (Borgstrom, 1980).

	Imports	Exports	Net imports	Net exports
Industrialised countries	18.7	9.4	9.3	-
Developing countries	1.7	10.8	-	9.1

The increases in efficiency of animal production that result from the new nutritional strategies will, if widely applied, increase animal productivity to the extent probably needed by the expanding human population in developing countries. However, it needs a retraining of technologists in these countries to absorb the new concepts and a move away from temperate country teaching of ruminant nutrition.

The tropical climate can be used to the advantage of the developing countries where ruminant production can be more efficient.

Acceptance of these feeding strategies could reduce the need for land clearing and pasture establishment in the fragile areas of the world which have been so prone to erosion following clearing. It may also allow for considerable change in the use of such pastures with reforestation as a highly desirable first option. In addition, reduced world ruminant populations would reduce incidental release of methane from decomposing dung and reduce the inputs of fossil fuels required in much of the infrastructure.

The Conservation of Animal Genetic Resources in the Developing Countries: a Practical Way Forward

The Economic Importance for Genetic Resources

Important economic gains may accrue from the appropriate choice of the livestock genetic resources correctly utilised in a given production system. This is illustrated by results a trial on dairy cattle crossbreeding strategies undertaken in Brazil, in which the accumulated lifetime performance records of 527 females have been recorded over the past 14 years on 67 farms. Six levels of crossbreeding between Holstein-Friesian (HF) and zebu Guzera (Z) were included with the HF genetic composition ranging from 1/4 to e" 31/32 HF fractions. A summary for the first 9 years.

Farms were grouped into two management classes for analytical purposes. The F1's were the most profitable group under both management regimes and for this reason was taken as a reference point to express relative performance of the other crossbreeding strategies.

Rotational crossing of HF sires for two generations followed by one generation of zebu sires (HF-HF-Z) was the second best alternative in the high management group. Upgrading to HF was as profitable as the HF-HF-Z rotation under high management, except under the higher fat and protein pricing which would be fairer to farmers. However in the low management grouping, upgrading to HF would have had disastrous consequences due to low production, high mortality and low culling rates. A new breed developed from *inter se* matings would require a high selection intensity to reach the same profitability as the F1. Choice of strategy would, of course, depend not only on performance but also on the actual possibilities of implementing each crossbreeding scheme (Madalena, 1989).

The above example shows that by using the right type of animal, without changes in either nutrition, health or other inputs, profit was considerably increased. Conversely, losses resulted from the inappropriate choice of crossbred. This, of course, does not mean that environmental factors should not be improved, but rather that both the genetic and environmental components should be considered in unison. The choice of germplasm is an integral element in the production system and it must be carefully matched to the other available inputs.

It might be appropriated to discuss the implication of the genotype x environmental interactions that are seen in the Brazilian trial. The view is sometimes expressed that there is no need to worry about genetics until the management is sufficiently improved to allow full expression of the existing available genetic potential. This view, however, fails to recognise that some genotypes have higher potential in a favourable environment but a lower potential in more stressful conditions. Therefore the notion that there is a genetic potential for each level of management is conceptually and practically more accurate (Falconer, 1960). As illustrated by the results shown ignoring genetic differences would be an unwise decision in any development programme.

Table 4: Profit per cow per day of herd-life under alternative strategies of crossbreeding of Holstein-Friesian (HF) x zebu (Z).

	Management Level*							
	"High"				"Low"			
Price-cost**	A	B	C	D	A	B	C	D
F1 profit***	1.8	2.8	3.5	2.5	4.6	5.5	5.1	5.4
Strategy	percent of F1 performance							
HF-HF-Z rotation	75	72	84	81	48	51	52	47
HF-Z rotation	41	53	63	57	59	60	61	63
Upgrading to HF	75	57	84	80	-21	-12	-10	-28
New breed (5/8HF:3/8Z)	-18	1	23	14	30	33	34	55

* Characteristics of management levels:	"High"	"Low"
Mean first lactation milk yield (kg)	2450	1731
Mean first lactation length (days)	283	309
Mean first calving interval (days)	402	561
Concentrates fed, kg/cow/day	4.5	1.6
All milked 2x/day with calf suckling to stimulate let-down.		

All milked 2x/day with calf suckling to stimulate let-down. ** A: 1980 to 1985 prices (protein not paid for) B: Fat differential tripled and protein paid at same rate. C: Cost of concentrates halved. D: Beef value of animals doubled. *** Profit expressed in kg milk/day. Price of 1 kg milk (3.3% fat) = US$ 0.16.

Source: Madalena et al. (1989).

Genetic Variation a Natural Resource

Historically, man has reshuffled genes from different livestock populations by crossing, selection and inbreeding. Biotechnology now offers powerful new methods to change the genetic composition of animals. However, because genetic material cannot be synthesized, improvements will still be restricted to the obtaining of the best possible combinations of existing DNA. Therefore, animal genetic resources constitute an indispensable natural resource must be properly managed for efficient production now and to be preserved for future use.

Although germplasm introduction has traditionally been a common practice in animal production, it was only really in the last two decades that genetic variation came to be viewed as a natural resource (Dickerson, 1969). Commercial use of genetic variation between populations has increased dramatically in some species as has research, primarily in developed countries, regarding the its theoretical aspects and the evaluation of alternative breeding plans.

Although migration also plays an important role in the genetics of animal populations in developing countries, research generally did not proceed, or even accompany, commercial trends. Genetic resources in these countries have not been adequately evaluated or fully utilised and in some cases are threatened with extinction without being properly described. The following activities might be listed for action on genetic resources management:

- description
- evaluation
- utilisation
- conservation.

FAO's programme has recently been described by Hodges (1990) and I will briefly refer to the above in the following sections, concentrating on issues rather than methodology.

Description

A brief description of the world's main livestock species and breeds is given in Mason's (1988) "World Dictionary of Livestock Breeds". A Global Data Bank of Animal Genetic resources has been established jointly by the European Association of Animal Production and FAO at the Institute of Animal Breeding and Genetics, Hanover

School of Veterinary Medicine (Simon, 1990). Information from the bank is publicly available. A questionnaire was developed to collect information on breed origin, numbers, phenotypic description, uses, management systems and relative performance compared to standard breeds. The work was initiated with cattle, buffalo, sheep, goats, pigs and horses, however, questionnaires for poultry and Andean cameloids are in preparation.

The databank currently has information on 658 breeds from 25 countries (mostly European, USSR and China) and procurement of data from other countries is being sought. In many cases such information is locally available although it is not generally readily accessible at hand, so compiling it in one databank requires motivation and effort, especially for completing the questionnaires. In other cases, special field surveys may be required to obtain the relevant information.

Estimates of genetic distance between breeds would be valuable in formulating conservation policies, thereby allowing a more rational selection of breeds requiring preservation. Where possible DNA description is preferable to indirect description involving gene expression (Primard, 1985). Trinity College, Dublin, is undertaking a study to measure genetic distances between 6 european and 8 tropical breeds (from India and Africa) of cattle. DNA finger-printing based on endonuclease restriction length fragments is being evaluated and both mitochondrial and nuclear DNA concentrates are being extracted locally and transferred to Dublin for analysis (D. McHugh and R. Loftus, personal communication).

Evaluation

The systematic introduction and evaluation of germplasm is a common practice in plant breeding but less so in animal breeding, except perhaps in poultry. However, important work has been undertaken such as the trials established by the Meat Animal Research Centre in Lincoln, Nebraska, where a large number of cattle breeds have been evaluated over the past 20 years; the early Argentinean beef breeds comparison trial in the 1960's; the evaluation of European breeds in Great Britain and the FAO trial in Poland to compare Holstein-Friesian strains from 10 different regions.

The evaluation of germplasm is just one particular aspect of the overall research requirements. One hesitates to emphasize germplasm evaluation since the benefits would appear to be obvious. Yet, there have been many cases where recommendations based on opinion, rather than of results, have had disastrous effects.

This is particularly the case where livestock importations have been involved and Vaccaro (1990) indicated that european dairy breeds are not sustainable.

Also the general belief that 5/8 European x 3/8 zebu cattle crosses were the best combination for tropical environments has caused unnecessary delays in new breed developments (Madalena, 1989) and, in spite of early warnings (eg. McDowell, 1972), it has only been possible to put the matter in its proper prospective in the light of recent experimental results.

Another example of the importance of germplasm evaluation may be found in the Criollo, cattle that were originally introduced into Latin America, and that have adapted over the centuries. The Criollo have now almost completely subsumed by the zebu in the lowland tropical areas, a process that intensified in the 1920's (de Alba, 1987). Recent research has shown that purebred zebus are no more productive than the purebred Criollos, although the F1 crosses are superior to both, due largely to heterosis expressed in reproductive efficiency and other economically important traits (Plasse, 1989).

Therefore, a crossbreeding policy would have been indicated instead of the breed substitution. As de Alba (1987) pointed out, farmers and ranchers impressed by the crossbred's performance attributed it to the new breed and not to heterosis - which is a more difficult phenomenon to grasp. The result was that a population of over 100 million head were pushed in the wrong genetic direction. Had the research been done 70 years ago and the results explained to farmers, this situation might have been avoided.

To be useful, germplasm evaluation must be comparative. The breeding alternatives should be compared over the same environments using appropriate designs, which include sufficient numbers of representative animals and sires.

Lifetime performance needs to be recorded, since both survival and herdlife are major components of the overall economic performance of breeds in stressful environments, and large differences in these traits are expected for diverging genotypes. With ruminants, this means that evaluations have to be in the order of 10 years. Shortcuts are not available and would only lead to underestimation of the real economic differences between alternatives.

Germplasm evaluation should, wherever possible, be conducted under commercial management rather than at experimental stations. In developing countries government farms are often inadequately funded, poorly administered and it may not be easy to simulate the variety of management and socioeconomic situations found in the private sector. For example, it is unlikely that administrators would allow the poor husbandry conditions found in some commercial farms. While the basic components of economic performance should be recorded on farms, some investigation of the more sophisticated traits may need to be undertaken to understand why the observed differences occur. For example, product quality may be conveniently analysed at a central laboratory, although getting the samples to it may be not an easy task.

Operationally, on-farm trials are usually cheaper, since farmers do not carry bureaucratic overheads. It is also less likely that administrative changes would disrupt long-term experiments in farms as can happen in experimental stations. On-farm recording, however, although not expensive, requires organisational and logistical support along with careful supervision.

I submit that the real opportunity cost of on-farm germplasm evaluation is not high. The costs of producing the necessary animals, distribution and performance recording may be largely recovered from production through some form of share-farming schemes as was done in the Brazilian trial. Staff and facilities for supervision, analysis and to interpret results would be available at research institutes/ universities in many countries, although they are likely to be overburdened with other activities. Therefore some support would be needed to strengthen laboratories, hire technical staff and provide minor, unsophisticated equipment and reagents.

Conservation

Characteristics of both species and breeds result from the expression of not only individual genes but also from gene combinations which interact to influence the physiological processes. Genes and gene combinations are renewed by reproduction, however, they may also be lost through either the extinction of the population or replacement by other genes or gene combinations. Selection and crossing will cause genetic change in a given direction, conversely, a low population size and inbreeding may result in the random loss of genetic material. Depleting genetic variation will restrict the choice of available genetic material for use by future generations. Should new circumstances arise, requiring animals with new characteristics then the genetic material would not be available to develop it. Conservation of animal genetic resources must be seen as one aspect of the wider problem of maintaining bio-diversity. Present increased resource utilisation conflicts with possible future needs.

As the decision making process becomes increasingly centralised through the development of rural organisations, communications, education and propaganda; and the increasing use of reproduction tools like artificial insemination and embryo transfer, so the power of man to change the genetic make-up of livestock will increase dramatically.

Cases are already known of breed replacement on almost a continental scale within a few decades. Some preservation is clearly needed to allow future generations the opportunity to be able to use the genetic resources that are currently available today. Thus, conservation of genetic resources could be considered as an insurance. Studies have shown that the economic return is very high when the preserved germplasm is extensively used (Smith, 1984). However, the likelihood of the preserved resources being useful in the future is, by definition, not known so the decision on how much to spend remains largely a subjective one.

At present, there are two basic ways of preserving genetic material for further utilisation, either as live animals or as deep frozen semen or embryos. Freezing of ruminant semen has been used for a long time and embryos for the past decade.

Freezing of pig and horse semen is a more recent development, although freezing pig embryos is not yet possible. Freezing of poultry semen is not yet commercially feasible. The conservation, or reactivation, of other cells, chromosomes or DNA should still be considered at an experimental stage (Brem and Brenig, 1990).

Because biotechnology is moving so fast, techniques may rapidly evolve, so it may be appropriate to set a conservation planning horizon of, say, 25 years. Future users would then decide if and how to reactivate stored material; continue to preserve it (or indeed to throw it away!) in the light of available techniques at the time. A short planning horizon is not reasonable, because most of the conservation cost is incurred initially, to establish herds or to freeze the semen/embryos.

Judgement is required to compromise between the size of the genetic sample, the number of unit to be preserved and the cost of preservation cost. Based on inbreeding considerations Smith (1984) suggested storing 100 semen doses each from 25 unrelated sires and 25 embryos from each of 25 unrelated females donors and sires per breed. Indicative costs of semen preservation for 25 years would be US$126,000 per breed and US$158,000 for embryo preservation. Sixty-five and 86 percent of these costs, respectively, correspond to the initial collection cost while maintenance costs are relatively inexpensive. Although semen is cheaper to preserve, reactivation would be faster if embryos were also available.

Alternatively, a herd/flock of 25 breeding females could be kept with no inbreeding over a 25 year period, provided controlled matings or artificial insemination was possible. Under pasture management this should not be expensive. Thus, in situ, conservation appears to be simpler than cryo-preservation. However, the risks of losing genetic material due to constraints in population size are higher for live animals. Disease outbreaks, droughts, broken fences, periods of poor management and the intrusion of unwanted sires in the herd might not be rare occurrences and would need to be guarded against.

Therefore, each conservation method has its merits and weaknesses which need to be assessed for each particular situation. For example, some countries have established embryo manipulation facilities and these could be made available to other countries in the region, such as, the proposal for Regional Gene Banks promoted by FAO.

In general, conservation of genetic resources is a long-term activity but there may be special cases in which its objectives may complement shorter-term development goals i.e. in the case of breeds that are commercially under-utilised, in spite of their potential to improve the economic efficiency of present production systems. Yet these valuable breeds may not be preferred by farmers or decision makers for a number of reasons, including: lack of research regarding their potential contribution, propaganda and/or vested interest. Germplasm from these breeds should be evaluated in the context of designing breeding strategies, including breed development, which would lead to their commercial conservation.

Development and conservation must go together. In fact, some contend that under-development leads to loss of natural resources (D. Wood, personal communication). There is also, however, a moral issue: it is difficult to think of saving endangered breeds when confronted with children suffering from hunger caused by famine. Yet, we do not wish to hand over to the next generation a world depleted of the majority of it's natural resources. Essentially mankind has the necessary genetic resources in terms of breeds and individuals to meet the challenge of feeding itself in the future (Cunningham, 1991). Some conservation effort is justified to sustain development in the long term and a balance will have to be established. It is certainly not logical to preserve resources for future use when we do not properly use them now. This would be like spending all your money insuring a car and then not be able to buy the petrol to run it.

Chapter 5

Practical Technologies and Options for the Genetic Improvement of Livestock in Developing Countries

Some of the implications for breeding strategies regarding the need to promote sustainable animal production systems in the developing world. Due to restrictions of space, cattle in the lowland tropics are taken as the main example and it is hoped that many of the same considerations will be applicable to other species. The choice seems justified on the grounds of the numerical importance of cattle in the tropics and the fact that so much more needs to be known about tropical lowland, compared with temperate, zone systems in all fields including animal breeding. Since such large areas of the lowland tropics are characterised by acute poverty and hunger that the case also well illustrates the need faced by developing countries to ensure immediate increases in cheaply priced food whilst, at the same time, conserving their natural resources.

National Objectives

Sustainable cattle production systems in the lowland tropics must contribute to the alleviation of poverty, hunger and national indebtedness in both the short and long term. Priority must be given to those systems which offer the best opportunities for providing cheap food and other products. This means that they will be based primarily on locally abundant feed, animal and human resources. The improvements proposed must also be widely applicable and not directed solely to an economic or technical elite, otherwise they will not contribute fully to rural development.

These restrictions immediately define certain biological aspects of the systems which should be given priority, as well as also important guidelines for genetic improvement. Of the biological aspects, the nutritive basis of the systems is perhaps the most important single feature because it determines the type of animal which must be used. To comply with the objectives outlined above, the base feeds involved will be those which are locally abundant.

Throughout the tropics, these will include agro-industrial by-products, crop residues and, in some regions, tropical grasslands. Grains, particularly imported cereals, will be restricted and, in general, concentrate supplementation reduced to the strategic use of specific nutrients to optimise ruminal function and the efficiency of use of the diet as a whole (Preston and Leng, 1987).

The use of by-products and grasslands through animals contributes to the sustainability of the whole production system. Furthermore, animals can use wastes which have little alternative use and simultaneously provide traction and fertiliser, if needed. They also contribute to sustainability in the direct economic sense by providing a source of low-risk savings and, if milked, of daily income (Winrock, 1978). Tropical Latin America alone has over 500 million hectares of grasslands, often occupying marginal acid soils or steep slopes (Sere and Jarvis, 1989). Grazing systems under these conditions have a key role to play in soil erosion control and fertility improvement, particularly when legumes are included. In addition, the improvement of grazing systems already established on cleared tropical forest should help to stem the rate of forest destruction by providing a better livelihood from the land which is already cleared.

These considerations have important implications from the genetic point of view. In the first place, whether by-products or tropical grasslands provide the basis of the diet, their nutritive value is low and the levels of individual production will not be high. Expressed in terms of milk yield, perhaps 4 to 10 kg/cow/day are reasonable limits. Yields much above this level will involve the use of resources (e.g. imported grains, heavy capital investments, advanced technical skill) which prevent the systems from meeting the socioeconomic objectives originally set out.

Thus, the genetic potential of the cattle for production must be carefully matched to the resources available and, in this respect, the temperate zone criteria are irrelevant. Secondly, all improvement measures, including genetic ones, must be simple and cheap enough to be widely applicable and the selection of appropriate measures must ensure that they make the best possible use of the resources involved. These same principles appear to be equally relevant to other animal species, whether ruminant or non-ruminant.

One of the most unsatisfactory features of pig and poultry production systems in many parts of their tropics is their total dependence on imported stock, technology and, often, feeds. This disregards the potential of small animals as contributors to rural development by exploiting their ability to make use of local foodstuffs, including wastes, and frequently represents an incomplete and unsustainable production system of which the feed base (e.g. imported cereals) is highly dependent on fossil fuels.

Breed Substitution

One possible strategy for genetic improvement is breed substitution. Defined as the introduction of commercial females, as well as males, this seems to be the least satisfactory option, if the strict socioeconomic objectives are borne in mind.

In the first place, if the constraints of the local environment are correctly assessed and the products required from the animals precisely determined, it may well be that existing genotypes are adequate and able to respond sufficiently and economically to improvements in the system. Frequently, the complexity of the required product is often overlooked. Pigs, for example, play an important role in many small farming systems as a low-cost and, therefore, low-risk source of savings. Once exotic genotypes are introduced, purchased inputs become essential and risks increase to a level at which the original objective is lost.

In the case of cattle, many lowland tropical systems require them to produce fuel, fertiliser and traction, besides meat and milk. Even where the first three are less important, as in most of Latin America, there are still strong arguments in favour of dual purpose, as opposed to specialised meat or milk production (Preston, 1976; Sere and Vaccaro, 1985). These multiple functions of cattle are an intrinsic element of the stability of the whole production system, which may well be upset if the existing stock are replaced. Resistance to disease and parasites may also be of vital importance. The trypanotolerance of West African N'Dama cattle is a case in point and increasing attention may be expected to be given in future to parasite resistance due to the high cost of insecticides and environmental concern. Tick resistance which is a characteristic of zebus or resistance to cattle flies such as *Dermatobia hominis* found in Latin American criollos (such as the Colombian Blanco Okejinegro) will also take on increasing economic importance.

Secondly, even where existing breeds have insufficient production potential, their substitution with new genotypes is, at least in the case of large species, a high-cost, high-risk measure which cannot be expected to solve the problems of the majority of farmers in a given locality. With cattle, neither substitution with another tropical genotype nor a temperate breed seems likely to be justifiable.

Few tropical breeds, or crosses, have sufficient numbers of genetically evaluated stock to justify importation, even presuming that their performance in the new environment could be predicted. Whatever their success, the possibility seems remote that the introduced animals will have any advantage over the crossbreds which could be derived from the existing population, except in terms of the time required to produce the crosses; which is, however, relative. Usually overall productivity of the existing animals is limited by poor reproduction and survival.

Investment in measures to improve these traits will have an important permanent impact on the efficiency of the production system and will pave the way for the higher potential crossbreds to be bred while the improvements are being organised. It would seem, therefore, that investment in environmental improvements along with crossbreeding is more likely to prove a cost effective alternative than the introduction of new tropical stock.

The introduction of European breeds into the lowland tropics is also unlikely to be a suitable option. Tropical forages are less digestible than temperate ones (Minson, 1980) and their utilisation therefore generates more metabolic heat. As European cattle do not have efficient heat dissipation mechanisms, high levels of concentrates are required if they are to survive and produce.

The concentrates are typically derived from cereals, often imported. These systems are clearly less sustainable in the long term than grazing systems or those based on by-products, for which pure European breeds are generally unsuitable. Furthermore, past evidence suggests that European dairy cattle will be unable to maintain their herd numbers, due to short and involuntary culling (Vaccaro, 1990). No genetic programme can be sustainable if it relies constantly on the importation of females. Besides, the costs involved in terms of concentrate feeds, veterinary supplies, technical skill and capital investment remove systems based on European cattle from the grasp of the ordinary farmer and prevent them from making the desired impact on rural development. Recent Latin American experience also questions their potential for providing low-cost food, since milk production costs have been found to be higher and profitability lower than with crossbred stock. It is extremely likely that the same arguments will apply to the introduction of exotic breeds of small ruminants and also, to some extent, to pigs.

Crossbreeding

A second option to be considered is crossbreeding. In the case of cattle, evidence from all over the lowland tropical world shows that, in one generation, tropical populations can increase yields to levels very close to the limits set by the locally available feed resources and which fit conveniently with the levels of technical skill, capital investment and service support infrastructure available.

Despite this, there are few examples of successful, stabilised crossbreeding schemes in commercial populations on a large scale, and many cases of patent failure - usually involving grades of European crosses above the levels which the local environment can sustain.

This situation cannot be attributed to any lack of discussion of the theoretical merits of different crossbreeding schemes for tropical cattle, at least in the context of milk production (Cunningham and Syrstad, 1987; Bondoc et al., 1989).

The difficulty appears to lie in the fact that the field success of crossbreeding schemes depends far more on their practical feasibility than on their theoretical merits. It is suggested therefore, that genuinely sustainable crossbreeding systems will not evolve unless the practical restrictions set by the local production systems are very carefully taken into account.

The more common of these practical restrictions deserve further examination. Systems which require the simultaneous use of more than one breed of bull per farm are difficult to put into practice unless there is sufficient infrastructure (fences, records) to separate one bull and his appropriate mates from the other. Although these difficulties could be overcome by changing the bull breed after a period of time, the problem of identifying sources of bulls of known genetic quality at reasonable prices remains. Extending the system to three or more breeds would seem to preclude its use on a wide scale in most practical circumstances. Besides, whatever advantages might be expected from additional heterosis are only of interest if there are no important additive genetic differences between the breeds available for the programme.

Table 1: Relative costs and profits derived from crossbred cattle of intermediate levels of European breed inheritance, compared with others under different systems in Latin America (50% European crosses = 100).

Country/System	Breed Group	Relative	
		Costs	Profit
Bolivia*			
Pasture + supplement	Local cross	100	100
	European	100	- 300
Confined	European	210	- 300
Brazil**			
Low level management	50% European-zebu	100	100
	25% European-zebu	40	40
	European	- 13	- 13
High level management	50% European-zebu	51	
	25% European-zebu	- 11	
	European	29	
Venezulela***			
Lowland, medium level management	50% European-zebu	100	100
	75% European-zebu	109	83
Highland, high level management	European	167	70

* Wilkins et al., 1979

** Madalena, 1989

*** Holmann, 1990

Thus, for example, rotational crossbreeding schemes involving Zebu and European breeds for milk must take into account the much lower yield to be expected from the Zebu bulls' daughters. Similarly, under Latin American conditions at least, it would be questionable whether a second European breed, besides the Holstein, would make a sufficient contribution to a dual-purpose crossing scheme involving zebu cattle to warrant the additional complication, due to the superiority of the Holstein with it most common rival the Brown Swiss (Syrstad, 1985). It is therefore not surprising, that the evidence available to McDowell (1985) suggested that no advantage has generally been observed from the addition of a third breed to crossbreeding schemes for milk production.

The practical difficulties involved in dealing with various breeds of bull on the farm could in theory be solved by using AI, although the difficulty of obtaining semen of good genetic quality remains. In this context, it is necessary to draw attention to the risks which are commonly associated with using AI for the production of routine pregnancies (as opposed to occasional, strategic uses). Before proposing any crossbreeding scheme which depends on AI, care must be taken to avoid the risk of losing more calves and lactations through lower conception rates compared with natural service, than can be compensated for by genetic improvement. This risk is typically high in lowland tropical cattle populations. From commercial dual purpose herds in Venezuela, Gonzales (1981) reported a 12% lower pregnancy rate using AI than with natural service and even under experiment station conditions, Paterson *et al.* (1983) reported a 22% difference between the two mating systems in South Africa. Differences of this magnitude are difficult to justify on the grounds of the genetic quality of the AI sires.

A second practical restriction refers to the lack of infrastructure for selecing local cows and bulls, which affects the potential success of *inter se* crossbreeding systems and of new "synthetic" breeds. Extremely few lowland tropical communities can take on successfully the challenge of being self-sufficient for genetic improvement, which is required once the population is closed. This probably explains why so few of the attempts to form new breeds in the tropics have succeeded. Cow selection requires more than recording schemes: data must be processed routinely to evaluate genetic merit. To this author's knowledge, no lowland tropical cow population is at present evaluated routinely for estimated genetic merit for production traits. The selection of bulls, through progeny testing, raises practical problems which will be discussed later, although it is relevant to point out that the problems associated with effective bull progeny testing represent one of the limitations of the *inter se* crossbreeding schemes.

Inter se crossbreeding programmes can also be criticised on the grounds that they are inflexible. Under Latin american conditions, a very small geographical area will include farms with widely different environmental conditions. Which would require animals of different levels of European breed blood and a fixed genotype is therefore of limited use.

These considerations lead to the conclusion that the most widely applicable crossbreeding schemes will allow: a) the use of natural service, b) the simultaneous use of one, or at most two, genotypes of bull per farm, c) the routine introduction of improved germplasm from outside, to bolster local selection efforts or even supplant them under extremely difficult circumstances, and d) the provision of different genotypes so that farmers, with suitable technical assistance, they may generate the type of crossbred which will be most productive under their specific conditions.

An option which may be of wider use is that of the using crossbred bulls, bred from selected local dams and proven European breed sires. Work is in progress to determine the optimum genetic merit of European sires to be used and, once this is clear, the possibility of giving tropical cattle populations access to the huge range of constantly improving temperate zone germplasm will open back. Local selection efforts to identify dams of potential bulls are required, but this can be done in pilot or nucleus herds without necessarily having to carry out recording and selection on the majority of commercial farms. It is assumed that a native (zebu or criollo) populations will be continuously available to provide bull dams using, perhaps, more marginal lands unsuitable for the crossbreds. If not, the scheme degenerates into a "grading up" programme.

Within Population Improvement

It is difficult to demonstrate whether the genetic progress to be expected from selection within an indigenous population is likely to justify the investment required. Heritabilities are seldom known with precision and selection intensity is frequently low under precarious economic conditions since sales tend to be determined more by immediate monetary needs than by biological criteria. On the other hand, the cost of obtaining, processing and interpreting records is likely to be high, because of poor and unreliable infrastructure, and equally difficult to predict.

Despite this, it seems unacceptable to propose that nothing should be done. Tropical cattle herds are extremely variable in production characters and whatever genetic basis for differences between individuals may be, effective management requires that superior cows should be retained and inferior ones culled. The need to get rid of unproductive animals is all the greater under conditions of restricted feed supply. Furthermore, farmers, certainly in Latin America, take great pride in the acquisition of bulls and regularly assign resources to this purpose. The opportunity for introducing improved germplasm into herds by this means should not be missed. Finally, in our experience, the process of encouraging farmers to assess their cattle individually and to take decisions as to their merit, is very well received and associated with spontaneous efforts on their part to introduce other improvements.

Table 2: The potential for selection in commercial dual purpose herds:the variation in production levels between the best and worst cows on five farms in Falcon, Venezuela

Farm No	Deviation from the group mean (kg)*			
	Milk yield/lactation		4 month calf weight	
	Highest	Lowest	Highest	Lowest
1	3058	-1694	52	-27
2	922	-861	38	-37
3	1175	-1096	42	-18
4	1484	-1391	26	-24
5	1634	-1215	39	-22

* 1584–2896 kg milk; 58-100 kg calf weight at 4 months.

Selection criteria and methods should be allowed to vary in complexity according to the stage of development of the programme. The important point is to start the scheme on sound, sufficiently simple principles, for it to be carried out properly. A step-wise approach for dual purpose cattle in Latin America which takes this evolution into account has been described elsewhere (Vaccaro and Vaccaro, 1989). At first, cow selection could be based on calf growth and/or milk and reproductive efficiency. The inclusion of some measure of fertility seems essential because of the evidence in zebu populations that heritability is moderate (Plasse, 1988) and the consistent evidence of a negative phenotypic correlation between fertility and milk production under tropical conditions. Where herd size is small, approximately valid contemporary groups can be made by uniting data across herds, classified according to production system and, possibly, mean production levels.

Table 3: Example of the negative relationship shown between milk yield and fertility in crossbred European x zebu cows in dual purpose systems in the lowland tropics of Venezuela.

Milk yield/ lactation (kg)	Interval from calving to conception	Anoestrus cows (%)
< 1000	68.1	17
1001 – 1500	92.6	22
1501 – 2000	104.7	28
2001 – 2500	121.6	42
2501 – 3000	137.4	57
> 3000	141.9	65

Source: Gonzalez (1980)

Bull selection for milk production poses special problems; to assume that they can be effectively evaluated by progeny testing may well be unrealistic. Tropical herds are characterised by low reproductive efficiency, late age at first calving and high mortalities, low culling and lack of identification. In addition, milk yield which is the trait of principal interest is extremely variable compared with temperate zone standards.

The size of the population which is mated by AI and performance recorded is usually small in the lowland tropics and, because of the high variation, low fertility and high rates of loss; more dams must be inseminated to produce the necessary number of daughters. This has been estimated at 30 in Cuba Menendez, (1985) and McDowell (1983) showed clearly the inaccuracies which result when daughter groups are smaller.

As a result, too few bulls can be tested reliably to permit an intensity of selection sufficient to justify the whole operation. In addition, the generation interval is prolonged by the late age for production of freezable semen and of first calving, as well as by the typical long process involved in processing and publishing the results. Once superior bulls are identified, their impact on the whole population is limited by the scope of the AI programme and low fertility associated with artificial breeding and by the high rates of loss before the end of the first lactation. An exercise to demonstrate the relative genetic and economic benefits of various selection options for dual purpose herds in Latin America showed little increase in genetic progress for milk due to progeny testing and a reduction in progress for 18 month body weight. It was concluded that the exorbitant cost of progeny testing under these conditions could not be justified.

Table 4: Simulated effects of different sire selection options on genetic progress in dual purpose herds under lowland tropical conditions in Latin America

Selection Method	Mean Generation Interval (yrs)	Cost/bull selected (Bs)	Calving %	Genetic Progress per year (kg)	
				Milk	18 mth body weight
Dam records (DR)	5.4	-	85	20.7	2.4
Sire records (SR)	5.4	-	85	56.8	2.4
DR + SR	5.4	-	85	76.8	2.4
DR + SR with embryo transplants	5.9	1,500	85	87.4	2.2
DR + SR with progeny testing of bull	8.8	341,600	65	89.5	1.5

Source: Vaccaro (1988)

Practical improvement programmes should therefore explore alternatives to progeny testing. In native populations, emphasis can be given to bull dam selection and in crossbred populations to sire selection as well if, as proposed above, bull sires are routinely selected from temperate zone breeds and used on native dams.

Where progeny testing can be carried out effectively, BLUP procedures are usually recommended. Few tropical populations are likely to have the required information and it is of interest to note that a comparison made in Cuba showed no difference in the ranking of 20 Holstein bulls whether the methods used were BLUP (with or without taking the between bull relationships into account), contemporary comparisons or least squares (Cordovi *et al.*, 1983).

Field performance recording is generally difficult under lowland tropical conditions, especially where herds are far apart and rains seasonally intense. Whole sections of various Latin American countries, including Peru and Colombia, are presently intransitable because of high personal safety risks. The best investment is probably to show farmers how to keep and use their own records. Wives and children will often spontaneously undertake record keeping and this source of enthusiasm should be tapped. Record processing may best be done locally. National organisations usually lack the stability and agility to be effective. Governments should therefore be encouraged to participate in the establishment of overall guidelines for record keeping and then delegate responsibility to properly trained scientists working locally in institutions such as universities.

The schemes should start on a small scale. The most usual bottlenecks are the time lag between record collection and return of the processed information to the farm, and the failure to present the results in a way which permits farmers to see the order of genetic merit of their animals for traits of priority importance. These common faults should be overcome before the programme is allowed to expand.

Given the difficulties set out above the establishment of pilot or nucleus herds appears essential. In extreme cases, recording and selection would be confined to these farms. The effects would filter down into the rest of the population at rates which vary according to their number and size.

Nucleus herds would generally be considered to be specific units but a very simple low cost scheme consists of providing suitable scientific advice to progressive commercial farms and using them as nucleus herds to provide at least some of the bulls and replacement females required by the rest. Care must be taken to admit only those farms which represent the priority production system, avoiding the temptation to use high yielding herds which do not.

Also, the farmers must be genuinely convinced that the selection criteria agreed upon are valid. The applicability of the proposal will vary between regions and will depend partly on herd size. The important principle is that official organisations in developing countries tend to be unstable and poorly budgeted, and there are clear advantages in locating the programme directly in farmers' hands.

One of the most important of the advantages is that it minimises the risk of selection under unrepresentative environmental conditions. The programme should then grow under its own momentum and be sustainable in the sends that it becomes increasingly independent of outside inputs. Such a scheme is currently in progress in Venezuela. Outstanding zebu cows are inseminated with semen from proven Holstein bulls and the young males distributed to surrounding farmers, who are kept informed of the number, genetic credentials and prices of the bulls available. The university base of the project is useful, at least in the first years of the programme, to give some assurance to purchasers that information is unbiased.

In other circumstances, it may be necessary to collect animals into one site to that proper recording and selection can be undertaken. For dairy cattle (or sheep) it might be the only practical way in which bull mothers can be mated to a few selected sires. If such a unit can be established on the basis of reasonably accurate estimation of breeding value, the genetic lift can be substantial. However, this may not be possible in all cases, and the major gain could simply be having adequate numbers to provide contemporary comparisons and the opportunity of recording objectively.

Screening a population to obtain the "best" animals is an important component of the establishment of a nucleus. Even where an existing herd is used as a nucleus, it is useful to sample the population outside albeit with the objective of only taking very few animals in. The principle of using contemporary comparisons can be maintained even though the group may be a village rather than each villager's herd.

There are clearly some criteria which can be used in all circumstances - contemporaneity and acceptability on a physical basis (if this is not so, dissemination may not take place). The criteria will range from proper direct measurements to stockmen's memory/judgement. If possible, it is useful in the early stages to obtain average animals from the same sources as the outstanding ones. Once together, the Comparison will provide a useful guide to the accuracy of the criteria and a direct indication to those interested of the value of genetic selection. Examples of this type of screening and nucleus formation are given by Timon (1990) and the results.

Table 5: Genetic Screening Results - Awassi Sheep Turkey

	Nucleus	Control	%
Lactation Yield (kg)	310 7.7	223 9.3	+ 3%
(range)	(254 – 469)	(97 – 360)	
Lactation Length (days)	206 1.7	187 4.4	+ 10
(range)	(159 – 224)	(95 – 222)	
Maximum Daily Yield	2.7 0.1	2.1 0.1	+ 29%
(range)	(2.1 – 4.2)	(1.0 – 4.3)	
No of animals	43	43	

Nucleus herds can allow selection to take place with limited training (in terms of numbers of people) since only the staff needed to run the unit and to record the necessary parameters need be trained. As long as proper monitoring is built in to such a system there is probably little need for a geneticist on site (except as a regular but infrequent visitor). While a central nucleus enables additional records to be taken (if cost effective), there is always the risk that the environment may not be representative of the sustainable production system or of a reasonable level of intensity within that system. The nucleus should never be provided with an environment better than that anticipated for commercial production two/three generations ahead. At least, such a constraint should allow the correlated response in commercial production to be within acceptable limits.

Open nucleus breeding schemes (usually known as ONBS) under most circumstances can achieve faster rates of genetic change although under certain circumstances (MLC, 1981) progress can be lower. However, the real advantages of the ONBS are the reduction in inbreeding and the fact that dissemination of the improved genetic material is built into the system. Such a fact also provides a good reason for achieving high health status in the nucleus but this can lead to major problems as stock transferred down the pyramid may well not be able to withstand the health challenges encountered.

Dissemination of Improved Stock

Perhaps one of the problems of all schemes in developing countries is the fact that if infrastructure is poor for recording it is equally poor for dissemination of genetic material. Certainly the derelict European style AI centres in Africa are witness to the problem. However, effective dissemination can be achieved using local resources as exemplified by the use of pig AI in some areas of China where local public transport and bicycles are the main forms of transport.

The problem of heat detection is always present where herds are of few animals and with buffalo, the problem is present in all herds unless a bull is used for this purpose. In cattle, the risk involved in the use of AI to obtain routine pregnancies under most lowland tropical conditions has been pointed out above. Usually, therefore, AI should be confined to use in specific cases, of which bull breeding would be one and the production of an initial crossbred generation perhaps another. Where natural service is required to ensure adequate birth rates, a great deal can be done to help farmers organise a reliable supply of males of known genetic quality through cooperatives or more informal networks.

Large quantities of resources are currently spent even in poor tropical countries on MOET. The excitement of new technology easily diverts scientists' (and politicians') attention from the strict socioeconomic objective of the production system. The possible benefit of alternative uses of the same resources (e.g. through AI) must be carefully measured before MOET is accepted as a viable option. The true genetic merit of the donor females must also be properly evaluated before transfer is carried out.

The difficulty of doing this properly should not be underestimated, especially for traits such as milk production which are of low heritability. It is possible that ET and also cloning could provide useful methods of multiplication. Where AI is likely to be difficult, it may be more feasible to provide "improved" male embryos for rearing within the locality so that distribution is achieved while still relying on natural service. Whether implantation is done centrally and females distributed or is done in the locality will depend on local circumstances. Again, however, it is essential to estimate the cost of the operation in terms of genetic gain and determine whether it does indeed make best possible use of existing resources.

Genetic Resource Conservation

It is widely accepted that genetic diversity is an essential element of the long term sustainability of production systems. It seems, however, unrealistic to propose that countries which now face severe problems of food production and poverty should devote their resources to the preservation of genotypes which are not at present commercially viable. Reduction in population numbers usually means that the economic potential of the animals is uncompetitive. It could also be argued that, given the wide diversity of natural ecosystems throughout the temperate and tropical world, a diversity of genotypes will automatically be maintained in production systems, at least in the case of species which are commonly relatively little protected from the natural environment (e.g. ruminants). In that case, it should be possible to obtain genes which might be required at a given moment, from populations which are maintained commercially in some ecosystems, even perhaps on another continent.

From these considerations it would appear, first, that endangered breeds in developing countries should not necessarily all be conserved. Realistic criteria must be established for deciding which cases are justifiable. These should include productive and adaptive aspects and the process could possibly be refined by genetic distance estimation procedures. It would also seem reasonable to propose that funds for this purpose should be obtained from international sources and that the work should be organised on a regional basis so that complementarity between individual nations' efforts can be improved. In considering conservation methods, the possibility of maintaining populations in natural parks or reserves should not be discounted although the risks of loss from disease and hunting must not be underestimated.

Finally, governments could reasonably be encouraged to take concrete measures to evaluate existing, small populations which could prove useful in the future in specific circumstances (e.g. in crossbreeding schemes or in traditional, small-farm production systems). One important element is to carry out the local research required to demonstrate the value of such animals in production systems considered most likely to be sustainable.

Planning Considerations

While it is relatively simple to comment with hindsight on programmes, the initial planning and development of schemes is frequently difficult. The difficulties often stem from the fact that different pressures are put on those involved in planning improvement programmes - political (governmental), breed societies/cooperatives, regional interests - and, in general, a belief that breed improvement is an automatic and acceptable solution. Since genetic improvement is relatively slow to provide change it can never be the "quick fix" so often cherished by politicians. The failures of so many schemes based on exotic semen are witness to that fact. The planner requires information on:

- the future production systems - feeds, management, housing (all based on the need for sustainability);
- the future market requirements and the likely infrastructure;
- the existing breeds and the population dynamics together with known performance.

Only then can long term, sustainable systems be planned and specific projects within the plan can be selected on the priorities agreed by those in the decision making role. This provides a more useful background for improvement since specific projects are known to fit into an overall picture rather than the more frequent occurrence of adopting a project and then trying to fit it into the long term strategy.

Policy Issues in Livestock Production in Arid Regions and the Management of Extensive Grazing Lands

Importance of Extensive Grazing Lands of The Arid Regions

Arid and semi-arid regions may be defined as areas where rainfall, relative to the level of evapotranspiration, is inadequate to sustain reliable crop production (eg Meigs, 1953) These areas are covered by grasslands, shrublands, savanna, semi-arid woodlands or desert. Kassas (1975) estimated that 43% of the world's surface is arid and Harrington (1981) suggested that in 1973 more than 40% of the world's population of sheep, 30% of goats and 25% of cattle were found in the arid zone.

Exploitation Systems

It is important to emphasise that the exploitation of extensive arid areas is concerned with land use and not the production of crops. Social organisation, ownership and access have profound effects on grazing management. There are four main systems of exploitation defined by the movements of the flocks and the extent to which the herders are sedentary.

Nomadic Systems: These systems are found in desert and desert fringes, where rainfall is extremely erratic and the flocks and their herders move to wherever forage is available, with no set seasonal patterns. True nomadic systems are becoming rare.

Transhumant Systems: Characterised by regular, seasonal movements of the flocks between grazing areas, often at different altitudes. In west Asia and north Africa flocks often move into the cultivated areas to utilise stubbles and by-products. The many changes that have occurred in transhumant systems recently are discussed below.

Semi-sedentary Systems: These systems are found mainly in Sub-Saharan Africa, where herders live in non-transportable dwellings but abandon them, if necessary, to move to other areas at times of feed shortage.

Sedentary Systems: Extensive grazing lands, where sedentary systems are practised, occur mainly in the developed countries of the

Americas, Australia and southern Africa, where properties have boundary fences and are often divided into fenced paddocks.

In other areas an important, but declining, area of extensive grazing land close to villages is exploited by the flocks of village farmers. Now, with the availability of trucks to transport water, sedentary systems are extending rapidly into large areas of extensive grazing land that formerly were only grazed seasonally because drinking water was not available for large periods of the year.

Private ownership of extensive grazing land is only found in South America, southern Africa and parts of the USA. In Australia, arid rangelands are owned by the State governments and in the USA parts of the range are owned by Government agencies.

The State, or Government agency, leases the land to graziers and, ultimately, they have the power to control the worst abuses of the land. Elsewhere, extensive grazing is generally an open-access or common-property resource, which may or may not have well defined regulations in relation to stock numbers and the duration and timing etc of their grazing. Regulation often existed on tribal lands or within tribal groups and, although it was often maintained under colonial rule, it has generally disappeared with independence.

It is widely assumed that degradation of extensive grazing areas is linked to over grazing and over exploitation of communally grazed areas, but it is clear that degradation has occurred and is occurring in areas where extensive grazing is owned or leased.

There have been problems on leased land in Australia (Harrington et al., 1990) and on freehold range in USA (Stoddart et al., 1970). On privately owned ranches in Patagonia and Argentina stocking rates are now 25–30% lower than 50 years ago because overstocking caused degradation of the natural vegetation and serious erosion (Mueller, personal communication).

Long-term sustainability of extensive grazing land can, therefore, be a problem under any system of land tenure.

The remainder of the paper will discuss the factors in ruminant production, predominantly sheep and goats, in west Asia and north Africa (WANA), which lead to the widespread degradation of the extensive grazing land in these areas.

Trends in Small Ruminant Production in the West Asia and North Africa Region

The recent and detailed analysis by Boutonnet (1989) of the Algerian ruminant livestock industry, which is dominated by small ruminants, has been taken as a starting point in this discussion. The number of ewes in Algeria has risen from 3.9 million in 1966 to 6.1 million in 1976 reaching 9.5 million in 1986. Seven million of these ewes, together with cattle, goats, horses and camels equivalent to 3 million ewes, are kept in the steppe.

Between 1971 and 1985, it is estimated that the carrying capacity was reduced by a half from 0.18 to 0.09 ewe equivalents per hectare. The 11 million hectare of steppe, therefore, can only provide about 10% of the feed requirements of the animals kept on it. Consequently, three quarters of the requirements of the sheep in the steppe are brought into the area and this necessitates the transport of 4.7 million tonnes of feed annually, comprising of 500,000 tonnes of barley, 400,000 tonnes of bran and 3,800,000 tonnes of straw. The remaining 15% of feed requirements is obtained by spring grazing in the desert and in the areas of cultivation after harvest.

The demand for red meat is high, mainly as a result of the rapidly increasing population (12.0 million to 23.0 million between 1966 and 1987), and a slight increase in consumption per head (eg 3.4 to 4.1 kg of sheep meat/head/year between 1970 and 1987), partly as a result of Urbanisation.

A key factor, however, in the expansion of sheep numbers has been the distortion of the market, created by regulations enforcing the compulsory sale of cereals to the Government, at fixed prices, while the market for red meat has remained free. Thus, the wholesale price of sheep carcase, adjusted for the cost of living index, increased from 38 DA/kg in 1970 to 102 DA/kg in 1987. In the same period the adjusted price for barley rose from 1.29 to 1.55 DA/kg. The ratio between the prices of sheep meat and barley widened from 30:1 to 66:1.

A further factor in this complex picture is the wide spread speculation in sheep. With a national flock of approximately 9 million ewes, Boutonnet estimates that nine million sales of ewes and ewe lambs occur annually, with large fluctuations in price. He states that the art of the flock owner in Algeria is less in the maximisation of meat production and more in the management of his capital of live animals.

Before independence, Algerian sheep production was characterised by periodic large fluctuations in sheep numbers with the affects of climate on feed availability. In bad years part of the flock was sold and the worst effects of drought on the steppe prevented. For 30 years since independence, the sheep population has steadily increased, as the state has always made sufficient imported barley available for the sheep numbers to be maintained in any particular year. Boutonnet quotes a peasant "in the past, in bad years, the shepherd sold part of his flock and kept only the sheep he could feed; now he sells his wife's jewellery to buy barley".

The increase in sheep population has been accompanied by a decrease in productivity. Carcase production per ewe and per year has declined as a result of a reduction in the number of sheep sold per ewe, in spite of an increase in individual carcase weight.

	Sheep sold/ewe	Carcase wtkg	Carcase wt/ewekg
1964-69	0.85	13.1	11.2
1974-79	0.72	13.4	9.7
1982-87	0.56	15.2	8.5

Boutonnet's description of the changes in Algeria and their causes has been presented in considerable detail, because it is derived from a careful analysis of a large amount of data and gives many insights into a complex situation. It emphasises the importance of the ratio between the prices of cereals and of sheep meat in stimulating flock expansion. In these circumstances, sheep enterprises, which in the past played a role as a reserve against uncertainty, have become a way of making speculative gains.

Boutonnet says "there is no need, in these circumstances, to invoke obscure reasons of tradition or socio-cultural prestige to understand the general propensity for an increase in the number of animals, since the economics of sheep production permit it, or for the lack of interest in improving flock productivity, in spite of the high price of meat".

A further point made in Boutonnet's analysis is that, although the rural population of Algeria has declined as a percentage of the total population, the actual rural population has not decreased because of the dramatic increase in total population. This strong and continuing pressure of rural people is in itself a contributory factor to the increase in sheep numbers. Where the ratio of meat to barley prices is favourable and there is access to grazing land, crop residues or even vegetable matter from urban waste, starting a small flock is a way of creating capital from a small initial investment. The large families mean that there is no shortage of children to shepherd the flock.

Boutonnet's report identified examples of this amongst farmers surveyed in the area of Sidi-Bel-Abbes. Small farmers with less than 20 ha of land, generally had no flock and sold surplus feed and by-products. If, however, they had a flock, the stocking rate was more than 10 ewes per hectare, far higher than could be supported by their land, and the flock was fed mainly on purchased feed. These flocks were generally financed from outside the farm, usually by some agreement with urban investors or landless shepherds.

Table 6: Total and Rural Populations in 8 Countries of north Africa and west Asia (FAO Production Yearbooks)

Country	Total Population (million)		Rural Pop. as % of total		Rural Population (million)	
	1975[1]	1987[2]	1975	1987	1975	1987
Morocco	17.3	23.3	54	38	9.3	9.1
Algeria	16.0	23.1	55	26	8.8	6.0
Tunisia	5.6	7.6	45	27	2.5	2.1
Libya	2.4	4.1	23	15	0.6	0.6
Jordan	2.6	2.9	30	7	0.8	0.2
Syria	7.4	11.2	49	26	3.7	2.9
Iraq	11.0	17.1	43	23	4.8	3.9
Turkey	40.0	52.6	61	51	24.5	24.8

= mean of 1974, 75 & 76

= mean of 1986, 78 & 88

FAO data for four countries in north Africa and four in west Asia and show that similar trends to those in Algeria are occurring in the other countries of the region. There are large increases in the human population in all these countries, except Jordan. The rural population, as a percentage of the total population, has declined in all these countries but the changes in the actual rural population vary considerably: no decline in Turkey, very slight in Morocco and small in all the other countries, except Jordan and Algeria. Sheep numbers have increased in all these countries, except Iraq and Turkey, in the latter, between 1975 and 1987 the numbers increased and then declined again. The FAO statistics appear to be too inaccurate to make any judgement as to whether the decrease in off-take of slaughter sheep and the increase in slaughter weight found in Algeria is a general trend. The area of barley cultivation and the production of barley increased in all countries except Libya and Jordan, both small producers.

Table 7: Area and Production of Barley and Sheep Population in 8 Countries of north Africa and west Asia

Country	Barley Area (1000 ha)		Barley production (1000 tons)		Sheep population (million)	
	1975[1]	1987[2]	1975	1987	1975	1987
Morocco	2026	2446	2279	2869	15.02	15.18
Algeria	826	1107	554	820	9.27	14.47
Tunisia	367	411	259	244	5.73	5.83
Libya	374	138	178	95	4.04	5.67
Jordan	57	37	22	27	0.71	1.12
Syria	960	1654	770	1509	5.87	12.55
Iraq	554	1243	516	1013	11.09	9.00
Turkey	2599	3314	4243	7133	40.67	40.12

1 = Mean of 1974, 75 and 76.

2 = Mean of 1986, 87 and 88

Syria - Degradation of the Steppe and Attempts at Prevention

In the last 15 years, major degradation has occurred in the Syrian steppe, where 75% of the sheep population is kept. This has happened in spite of the adoption of a range of measures intended to prevent continued over-utilisation.

In the mid 1970s it was possible to describe (eg Nygaard *et al.*, 1982) the traditional seasonal movements of flocks from the steppe to the agricultural land in June to graze stubbles and other byproducts. The flocks then returned to the steppe from November until the following June. Movements still occur but they are now less regular and, when they return to the steppe, the flocks stay in a small area around the family's house and are fed purchased feed.

A number of factors have contributed to the degradation. Masri (1979) suggests that a key factor in starting the process of degradation was the introduction of tractors in the 1950's, which allowed easy cultivation of the steppe, with the areas with the best rainfall and soils being cultivated first. Later, in the 1960's, drilling of new wells by the Government allowed grazing in areas which were formerly protected from grazing for long periods by lack of water. Then the purchase of water tankers permitted richer flock owners, with capital, to graze the remoter areas for longer periods. Before, these remote areas were a reserve, not because of tribal tradition or agreement, but because of their isolation and lack of regular watering facilities. Pick-ups enabled large areas to be covered looking for good grazing and the flocks could be moved rapidly by lorry. Thus the pressure on the remaining steppe increased. This was further intensified by the progressive abandoning of the tribal control of range land and the permitting of areas to be transferred to individual owners, which is still continuing.

From about 1960, the Syrian Government introduced a series of measures, which experts at the time thought would prevent further decline of the steppe. About 50 sheep and range cooperatives were established between 1968 and 1978 to provide credit to nomadic flock owners for the purchase of feed and livestock, with the aim of reducing fluctuations in stock numbers. The survey by Nygaard *et al.* in 1982 makes almost no reference to the improvement of range, which was initially a major objective of these co-operatives. To help decrease the grazing pressure in the steppe, about 50 fattening co-operatives were also set up to take light lambs from steppe flocks and fatten them on cereals. Membership of these co-operatives gave preferential access to feeds at controlled prices from the Government purchasing organisation.

The results of these initiatives have not been, as was intended, a stabilisation of the fluctuations in sheep numbers, and in the cycle of low prices for sheep meat in dry years and high ones in good years. Instead it has resulted in a steady expansion in sheep numbers since about 1970. This has led to a rising demand for barley, although the ratios between the prices of sheep meat and barley have been much less favourable than those in Algeria. For example, in 1988 the ratio was 15:1 with the free market price of barley and 18:1 with the controlled market price, and in 1989, when drought increased sales of sheep and depressed the price, the rations fell to 9:1 and 10:1 respectively. The demand for barley has caused a steep rise in the area planted to barley (Fig 1), which further intensifies the grazing pressure in the decreased area of steppe. The palatable bushes are over-grazed and disappear, and the unpalatable ones are often removed for fuel, especially in settled areas, which now extend far into the steppe. Even unpalatable bushes are important for the survival of the degraded steppe, as they help stabilise the soil and offer some protection from grazing for a few annual plants until they can flower and set seed. Further rapid degradation and complete loss of bush cover results from speculative barley growing financed by urban investors and which now extends into areas with less than 200 mm of rainfall.

Policy Issues

This description of the situation and changes in the west Asia and north Africa region indicates the enormous scale of over-grazing and environmental degradation in the region. The only possibility of achieving long-term sustainability in animal production systems is for governments to tackle the social and economic situation of the flock owners and farmers in these areas, so that it becomes economic for them to adopt more sustainable systems. It is almost impossible to believe that this can be done by any government, whether democratic or not, before further major damage to the environment occurs. This is mainly because the pressure of rapid population increase is the major determining factor and this cannot be altered in the foreseeable future.

Other policy issues, which could have some effect in reducing problems of over-grazing are discussed below.

Limiting Rural population

Rural depopulation and the availability of alternative employment will undoubtably have some effect in reducing the agricultural population towards a level that the land can sustain. In general, however, in these areas, rural populations are not declining quickly and some are still increasing. If stocking rates are to be reduced to levels that are sustainable in the long term, governments will have to introduce strong legislation and strictly enforce it. Again, it seems unlikely that this will happen: in the short term such legislation would cause enormous hardship to one of the poorest sections of society and its introduction would require either the provision of alternative employment or compensation for loss of a traditional way of making a living.

Pricing Policy for Inputs

It is clear that low prices for purchased feed are a major cause of flock expansion, because flock owners make rational economic responses. Government policy should be directed to eliminating subsidies and market distortions. To be effective, it must extend to bread flour, as subsidised bread becomes an animal feed when the price ratios are right. Fuel pricing policy must also be considered, as low diesel prices reduce the costs of cultivation and of the transportation of feed, animals and water.

Investment Policy

It is clear that sheep production is a good investment in many of these arid, developing countries, where there may be very limited ways for private investment. Speculation, however, in barley growing in the Syrian steppe, financed from the towns, is a serious cause of degradation. Availability of alternative investments in countries with poor financial services, might help to divert speculative investment from these forms of environmentally hazardous production. In the Islamic countries, where the gaining of interest on loans is forbidden by the sharia, there are particular difficulties.

Chapter 6

Technical and Research Considerations

Although change in socioeconomic factors is undoubtedly necessary to limit degradation, research has a clear contribution to make in defining problems and providing solutions, where and when socioeconomic conditions suggest that technical change may have an effect.

Some of the topics that require research are set out below.

Socioeconomic Studies

Research, similar to Boutonnet's study of Algeria, is needed to define the socioeconomic situation in each country. This needs to make full analyses of feed supply from all sources, including any unofficial market and that grown illegally. It is also important to quantify trends in off-takes of slaughter animals to establish whether systems are becoming less efficient in most countries.

Research is needed into the patterns of rights of access to rangeland: even if traditional systems have disappeared or been abandoned, it is too simplistic to assume that access is entirely free. The current situation, after a long period of ploughing in the steppes, should be described. It is clear that planting a barley crop establishes rights to the crop in its harvest year, but does it establish rights to the land subsequently? ICARDA is actively working on this in Syria.

Farmer Participatory Research (FPR)

Farmer participatory research is needed to define the farmers' view of their problems and discuss with them possible technical solutions. I see FPR as the only way to tackle the problem of low off-take of slaughter animals, which may be an increasing problem. Very close contact with flock owners is needed to identify specific causes and seek solutions.

Improvement of Extensive Grazing Land

The possibilities for improving productivity and sustainability of rangeland need investigating. Because productivity is limited by the amount of rainfall, technologies requiring costly inputs will never be economically viable in these arid areas. Options do exist and ICARDA has programmes on these for the WANA region.

In order of increasing input requirements, they are:

Improved grazing management. In areas with annual plants, management directed to improving seed set will slowly increase productivity. Currently, these plant communities are grazed heavily in late winter, limiting early growth and making very little contribution to the animals' nutrient requirements. Lower grazing pressures and the deferring of intense grazing would slowly allow an increase in herbage production. For areas where the imbalance between animal numbers and pasture resources is not too great, guidelines for grazing management have to be established for flocks that are continuously shepherded. Traditionally high grazing pressures (animals/unit area/ unit time) are used to achieve very high utilisation of the herbage available at a particular time. Continuous stocking systems are very difficult, if not impossible, to introduce for shepherded flocks.

The benefits of deferring the start of heavy grazing in early spring needs demonstrating, especially when shown that flocks walking long distances to graze very sparse pasture actually require more supplementation than continuously penned animals. Thompson (personal communication) showed that a flock taken to graze for 6 hours daily on poor pasture required 2.2 kg/head of supplement to maintain weight, while a continuously penned flock only required 1.7kg/head.

Application of fertilizer. In many soils in the region low soil phosphorous (P) limits growth and small applications of P in the presence of legumes can rapidly increase herbage production, if the grazing pressure is controlled. For example, legume seed numbers increased from 3.1 to 18.6 thousand per m^2 over 4 years with annual applications of 25 kg of P_2O_5 (Osman *et al*, 1990). Stocking rate was 2.5 times that of the unfertilized land.

Introduction of Legumes. Cocks (personal communication) suggests that, in Mediterranean areas, very small-seeded legumes of genera, such as *Astragalus* and *Trigonella*, may warrant research for areas with rainfall of about 250 mm. Species from these genera possess reproductive strategies adapted to these dry areas: very small seeds, that pass largely undigested through animals, large numbers of seeds per pod and high levels of hard seededness.

Introduction of Edible Shrubs and Trees. Shrubs, such as *Salsola vermiculata* and *Artemisia*, were an important component of natural steppe, not only as feed sources, but also to stabilise soils. Much research has been done on the establishment of plantations of bushes, particularly *Atriplex* species, both native and introduced, but there is a lack of long-term data on carrying capacity and animal performance. Although wide spread growing outside government stations is not common at present, they probably offer the only possibility for regeneration of degraded areas of steppe, particularly as low cost establishment from seed shows considerable promise.

Practical Technologies to Optimise Feed Resource Utilisation in Reference to the Needs of Animal Agriculture in Developing Countries

Production from a herd or flock of ruminants is a result of the interactions of environment, the animals nutrition, and its genotype.

Individual productivity from most ruminants in all developing countries is low. The reasons for this are complex but in order of priority appear to be:

- the imbalance of nutrients that arise from digestion of the available forage resources when these are fed without supplements,
- the incidence of disease/parasitism, and
- the often harsh climatic conditions.

Genotype is over-emphasised as a primary constraint as poor nutrition has an overwhelming effect. Resistance to disease and high temperatures of particular breeds is however an important overall consideration.

Recent nutritional research has demonstrated the possibility of very large increases in animal production that can be achieved by small alterations to the feed base. As these increases have also been achieved at the farm level without alteration to the other management practices, it demonstrates the large impact potential of the feeding strategies in the present environment. Increased production of meat/

milk with better body condition of animals also lifts lifetime reproduction rates. The lowering of the age at puberty of heifers and a decrease in the intercalving interval in cows have probably the greatest effect on the overall level of production.

Increased efficiency of utilisation of forage by the animals together with improved reproduction rates have demonstrated that production can be increased by up to five fold without changing the basal feed resources. This has been achieved, by providing the critical catalytic nutrients that are deficient in the diets and by balancing availability of nutrients closer to requirements. In general, the supplements required are urea/minerals and a source of bypass protein. In many countries these supplements are already available, in others, there is a need to manipulate these resources, to provide them in the correct amounts and in the appropriate form.

In many grazing areas the basic resources are not available locally. Research and development is needed to produce them economically at the centres of ruminant population densities. In the rangelands, particularly in the semi arid areas, tree forages, seeds and pods represent by far the greatest potential source of protein meals.

Available Feed Resources for Ruminants in Developing Countries

Throughout the last thirty years the expansion of crop and livestock production in developing countries have more than doubled but the increase in demand for food has been even greater, leading to an increase of food imports by approximately 10% per year. This situation is expected to remain at the same rate for the foreseeable future.

The increased production of animal products in developing countries has been largely through increased animal numbers, while production per animal has been static or increased to a minor extent over a long period of time (Jackson, 1981). The demand for food for humans in these countries is likely to increase and, since cropping land is almost fully utilised at the present time, it appears that cultivated pasture is likely to become scarcer in most parts of the developing world.

The ruminants niche is likely to remain as a utiliser of carbohydrate biomass which is not digested by intestinal enzymes and, therefore requires fermentative digestion, and which cannot be used extensively by monogastric animals (i.e. forage, crop residues etc.). The ability of ruminants to convert otherwise waste biomass into meat, milk and other products and to accomplish work suggests that they will endure into the foreseeable future. The ever increasing pressure on land for crop production suggests, however, that they will have to continue to do this utilising crop residues, industrial by-products and pastures from relatively infertile rangelands. The common characteristics of such feeds are low digestibility, low protein content and a low mineral component.

Animal Productivity

Animal Productivity from Available Feed Resources: Research over the last twenty years clearly indicates that it has been a popular misconception that low productivity of ruminants in developing countries is a result of low energy density of the available forages (i.e. low digestibility). This concept, often repeated in reviews, even up to the present time, is misleading.

There is now abundant evidence that low productivity stems from an inefficient utilisation of the feed because of deficiencies in the diet. These deficiencies are of nutrients critical to the well being of the microbes which ferment or digest the feed, and nutrients required to balance the products of digestion to requirements. This has been considered in detail recently by Leng (1991), who suggested that because an inefficient utilisation of nutrients increases metabolic heat, that the often low intakes of poor quality forage of ruminants in the tropics (where most developing countries exist) is imposed by a combination of climatic and metabolic heat stress (Leng, 1989).

Correction of a nutrient imbalance by feeding a bypass protein often (but not always) increases the intake of poor quality forages to a constant intake of around 80-100 $g/kg^{0.75}/d$. Tropical conditions impose this particular feed intake constraint on ruminants for a considerable period of a feed year and it is rarely seen in temperate areas.

On the other hand cattle in the tropics require less feed for maintenance, if they do not have to combat cold stress, and if they can process these extra nutrients they can be more efficient than animals on the same feed in a cold climate. To process the extra nutrients however, they need extra protein and thus the requirements for amino acids are higher in cattle in the tropics then in animals on the same feed in cold environments.

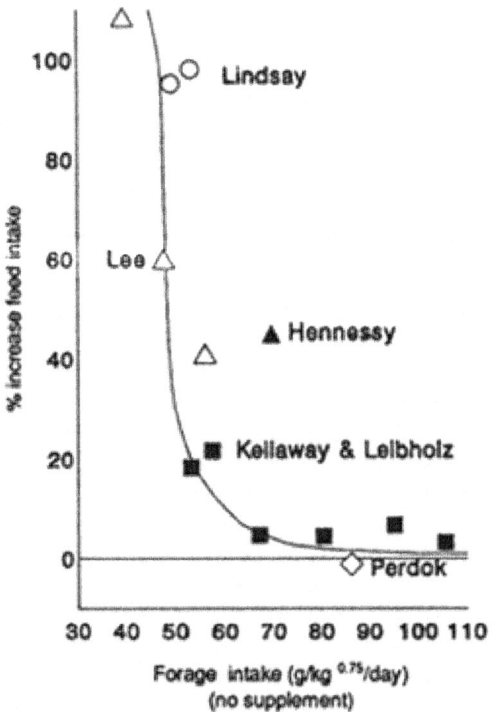

Figure 1: Intake of low digestibility forages by cattle either unsupplemented or supplemented with bypass protein or bypass protein and urea (Lindsay and Loxton, 1981; Lindsay *et al.*, 1982; Hennessy, 1984, Perdok, 1987; Kellaway and Leibholz, 1981).

Productivity Levels Achievable by Ruminants on Low Quality Forage Based Diets

The rationale and concepts on which the following discussions will be centred have been reviewed by Preston and Leng (1987) and Leng (1989,1991). The basic concepts are as follows. Ruminants fed low quality forages require supplementation with critically deficient nutrients to optimise productivity. The supplements that are required:

- correct any nutrient deficiencies for the rumen microbes, and
- they balance the ratios of protein (absorbed amino acid) to the energy (VFA) that arises from the fermentative digestion in the rumen so that it corresponds to the animals requirement.

The nature of anaerobic microbial fermentation in the rumen indicates that microbial cell growth in the rumen (which supplies the protein {P} to the animal) relative to the production of VFAs (E) (the major source of oxidisable substrate for ATP generation), will be extremely low in nutrient deficient rumen medium or digesta.

This means that a sub-optimal level of any nutrient for microbial growth in the rumen will result in low protein to energy (P/E) ratio in the nutrients absorbed. Ensuring a nutrient non-limited microbial digestion in the rumen by supplementation automatically improves the P/E ratio in the nutrients available to the animal. Feeding a protein meal in which the protein has been made insoluble or otherwise not attacked by rumen microbes, is a further major method for adjusting the P/E ratio upwards.

The ratio of microbial proteins to VFA produced and the effects of supplementation in a steer given 4 kg of organic matter which is completely digested in the rumen for a number of feeding conditions.

The reason for discussing these theoretical calculations at this point is to emphasise the large differences in protein to energy (from 12 to 50) in the nutrients absorbed by ruminants, fed unbalanced diets and diets balanced with supplements.

It also emphasises that on a diet high in bypass protein the rumen microbes need not be highly efficient. To obtain the higher P/E ratio without supplementation, however, to directly improve rumen condition, more bypass protein is needed.

Table 1. The effects on P/E ratio in the nutrients absorbed of supplementation with a bypass protein to cattle with a poor or optimised (i.e. supplemented) microbial milieu in the rumen. The values are calculated for a steer digesting 4 kg DM in the rumen.

Rumen environment protein (g/d)	Protein bypass (g prot/d)	Microbial cells Produced	Protein microbial (g/d)	VFA produced (MJ)	P/E* (g.MJ)
Poor	0	830	500	41	12
Optimised	0	1680	1010	30	33
Poor**	400	830	500	41	22
Optimised	400	1680	1010	30	47

* Microbial protein plus dietary protein to VFA energy.

** Although the rumen environment is deemed not to change through the addition of protein meal, in fact it will have been improved but may not be optimised to the extent it would by feeding a molasses/urea block. P/E ratio here is underestimated.

It is the relationship of P/E with the efficiency of feed utilisation that has a very large effect on growth, milk yield and reproductive performance.

The levels of production achieved when P/E is increased have been greatly superior to that predicted from present day feeding standards based on the metabolisable energy of a feed.

Feeding Standards and Feed Evaluation

Most forages consumed by livestock in developing countries have a low digestibility which rarely exceeds 55% and is mostly in the range of 40-45%. The calculated metabolisable energy in the dry matter (M/D) thus ranges from 7.5 down to 4.8. Feeding standards indicate that feeds with a metabolisable energy content of 7.5 will support growth rates of cattle of approximately 2 g/MJ of M/E intake. On a forage at the lowest level of ME, cattle would be in negative energy balance.

The relationships found in practice with cattle fed on straw or ammoniated straw with increasing level of supplementation in Australia, Thailand (") (Wanapat et al., 1986) and Bangladesh (Saadullah, 1984). Recent relationships developed for cattle fed silages supplemented with fish proteins (Olafsson and Gudmundsson, 1990)(Ä) and tropical pastures supplemented with cottonseed meal (Godoy and Chicco, 1990)(*) are also shown. This illustrates the marked differences that result when supplements high in protein are given to cattle on diets of low ME/kg DM.

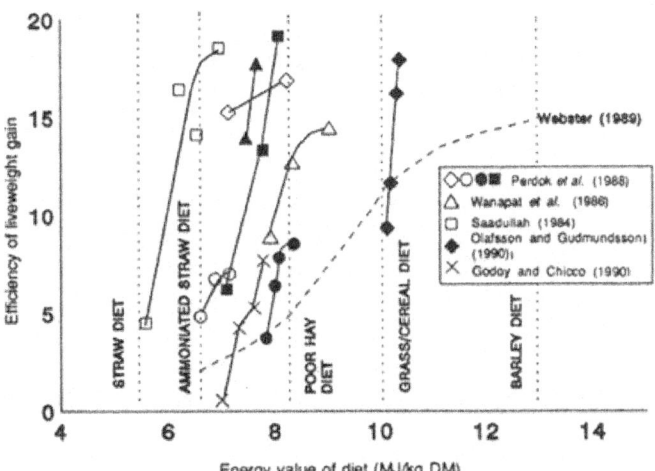

Figure 2: Schematic relationship between diet quality (metabolisable energy/kg dry matter) and food conversion efficiency (g liveweight gain/MJ ME) from Webster, 1989.

Contrast this with results of supplementary feeding trials based on balancing the nutrition of animals with urea/minerals and bypass protein, where cattle growth rates equivalent to 18 g/MJ of M/E intake have been achieved in cattle fed straw.

Obviously the presently accepted feeding standards have been very misleading and can not be used as a means of predicting animal performance. Of vital importance however, is that the application of the concept of balanced nutrition can improve animal growth by 2–3 fold and the efficiency of animal growth by as much as six fold over previous estimates (a range of 2–10 fold). In addition it also shows that although growth rates of cattle are below those on grain based diets cattle on forage based diets can be as efficient in converting feed to liveweight gain.

Low productivity of ruminant livestock has been accepted in developing countries as an inevitable result of the poor feed base and a low feed conversion efficiency. The concept being that there is a large heat production (energy requirement) associated with the ingestion and movement of digesta along the tract in animals fed on forages as compared to concentrates. This conclusion is clearly contrary to the conclusions of Leng (1991) and the concept of balanced nutrition presented here.

However, poor growth rates are associated with a slow maturity and, on poor quality roughage diets, cattle reach puberty at four to five years of age and often have a calving interval of 2 years. Recent observations on balancing the nutrition of cattle on such forages indicate that it is possible to reduce age at puberty by one to two years and potentially possible to support a calf every year on these same feed resources. Management considerations, however, suggested that 15 months calving interval is more likely. The flow-on effects of improved reproduction are: increased percentage of a herd in production and an increased offtake of animals. The flow-on from improved reproduction outstrips the direct effects on immediate liveweight gain or milk yield.

Application of Supplementation to Balance P/E Ratios

Even though the principles of feeding bypass protein to improve productivity have been known for many years, application has been slow and unspectacular, particularly in the developing countries. The application has been slowed by:

- the inability of research scientists to communicate and be believed by applied technologists;
- the desire by many scientists to stand by the feeding standards that have been promulgated for twenty years and which appear to be totally inappropriate for most feeding systems;
- the controversies surrounding the principle mechanisms of action of protein supplementation which has clouded the major issues.

However, wide scale application of the use of supplements high in bypass protein have occurred in India through the initiatives of The National Dairy Board of India (NDDB). For the same reasons as given above, progress was initially slow (development started in 1980) but it is now accelerating at a pace which should see most feed mills in India dedicated to the production of bypass protein supplements in the next five years. At the present time approximately 100 MT of bypass protein feed is being fed daily to their dairy animals by small village farmers. In many situations, this is coupled with the use of a molasses urea block, particularly by the more advanced farmers.

After considerable experimentation and village testing of a bypass protein supplements, the management of the cooperative feed mill in Kedah district in India decided to convert from the production of concentrates based on traditional concepts to a bypass protein concentrate (30% CP) composed of locally available protein meals plus 10% grain and approximately 10% molasses. After pelleting, the protein in the concentrate was found to be 75% insoluble in buffer solution (Leng and Kunju, 1988)

The marketing strategy was to widely advertise the new feed concentrate with a simple statement that village farmers should "feed to their dairy animals half the weight of the usual concentrate and this would double milk production". As the feed was about a third more expensive, there was some considerable opposition to its introduction. Nevertheless, the feed mill (capacity 100 MT/day) was converted to the production of new feeds on December 1st 1988. The feed was purchased, somewhat reluctantly by some of the farmers, but all opposition to its supply appeared to have been overcome by the middle of the next hot dry period (April-June) when milk yields were considerably above previous years with only half the weight of supplement.

The collection of milk within the dairy co-operative, relative to the feeding of the traditional and new feed supplement for the previous five years and for the twelve months since conversion to the new feeding systems. Whilst a number of changes have occurred in the area which could contribute to the increased milk collected, from the research carried out by NDDB prior to the change over the responses in milk yield are in line with observed increases in milk collected.

The effects of the changed feeding appear to be a 30–50% increase in milk production, from Dec. 1st. 1988 to Dec. 1st. 1989 - a further similar increase in production is apparent in 1989-1990 (unpublished observation). The further increase probably represents the flow-on effect that would come from increased reproduction rate that should have resulted from the new feeding systems.

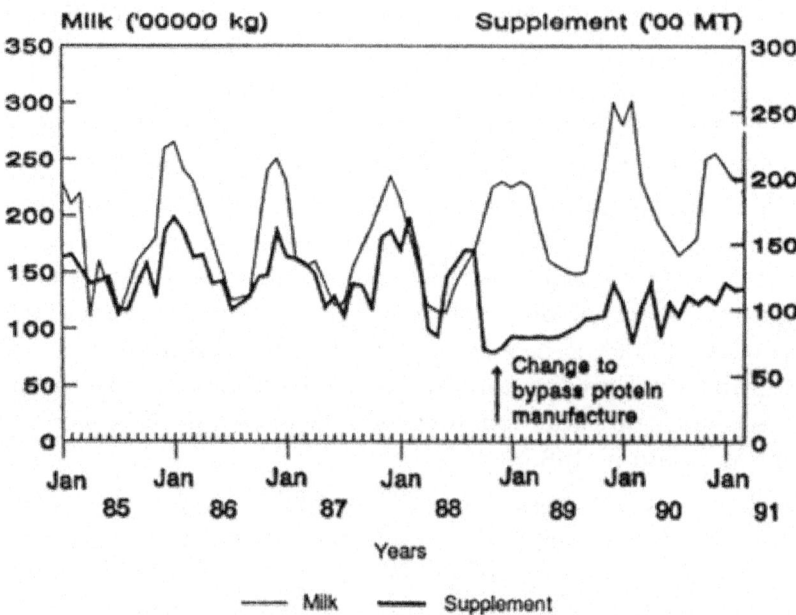

Figure 3: Milk collection records and the sale of supplements in for the Co-operative in the Kedah district of India when supplements were compounded on traditional concepts (1985–1987) and following (1st December, 1988) their replacement with a 30% C.P. bypass protein pellet (records provided by the NDDB of India).

Impact of New Feeding Systems on Milch Herd Composition

Undoubtedly the new feeding systems, combined with disease control and better management, is facilitating the introduction of cows with a higher genetic potential for milk production. The apparent lowering of heat stress through the balanced approach to nutrition has also allowed Friesians (the mothers of bulls that will be used for cross breeding purposes) and selected buffaloes, to yield milk well above the average reported for the developed countries. The yields of imported Friesian cows from Germany fed on all roughage diets supplemented with a molasses/urea block and a bypass protein (recommended at 350 g/litre of milk) at Anand averaged over 5,000 litres/305 day lactation in 1988–89 with individual milk yields of over 6,000 litres/305 day.

These milk yields are achieved in an area of India considered to have one of it's hottest climates. Higher yields were observed with Friesians managed by village farmers in a more temperate area of India (Bangalore).

Practical Methods for Balancing Nutrition of Ruminants Fed Low Quality Forages by Providing Deficient Minerals

Provision of Minerals

It is not practical to identify the major critical micro and macro minerals in a basal roughage diet, as these will vary from site to site and year to year. The practical approach is one of 'rules of thumb' that provide a best bet or 'shot-gun mixture' of minerals as economically as possible. A concentrated plant extract, such as, molasses provides such mixtures and can be fortified for specific areas where local knowledge points to specific deficiencies. In this respect, molasses (both sugar cane and beet) and concentrated palm oil sludge offer useful sources of these minerals. They are also quite palatable to livestock and are useful in hiding less palatable nutrient sources in supplements.

Mineral salt mixtures are commercially available in most countries. They usually have a high content of salt and only minor quantities of the trace elements. In practice, fortified materials, such as molasses, will be superior to commercial mixtures as they present a greater coverage of all the minerals required, also thy are a valuable source of other nutrients (e.g. B vitamins) and a small amount fermentable energy.

Supplying the Rumen Microbes with Ammonia/Urea

The other requirement is for a non-protein nitrogen source for the rumen microbes, usually urea. Urea is usually administered together with the minerals and its concentration in such mixtures is controlled by safety and ease of incorporation and, therefore, rarely exceeds 10–15% of such mixtures. However, this is usually sufficient to allow an intake of between 50 and 100 g of urea by cattle from a molasses/urea block which is sufficient to balance ammonia in the rumen of cattle on a low N roughage diet.

Results from India suggest that mineral/urea mixtures in, for example, molasses multi-nutrient blocks, are best given *ad libitum* allowing the animal some degree of selection and there are indications that the animal will learn to control the intake of urea to an optimal level.

Bypass Protein Supplements

Providing bypass protein to cattle under small farmer management is often difficult and, at times, is too expensive. There is often little information on the locally available protein sources, particularly the level of protection of the protein in the rumen. As a rule-of-thumb, solvent extracted oilseed cakes, fish meal that has been flame dried (but not sundried or fish silage) and protein sources that have been heat treated, have considerable protection from rumen degradation. The degree of protection is enhanced by pelleting the protein meal in the presence of free glucose or fructose, (as occurs in molasses) when a mild browning reaction occurs (unpublished observations).

The turning over to supply bypass protein rather than concentrate for supplementing the cattle of small farmers, the National Dairy Development Board of India converted the existing feed mills to the production of high protein supplements. Wherever possible, solvent extracted oilseed meals form the basis of the supplement. However, the protein ingredients have to be purchased on the open market and leastcost formulation is desirable because of the size of the production unit (50–100 MT/d.). The high degree of protection of these protein feeds is achieved by ensuring a high proportion of solvent extracted meal and also by heat pelleting with 8% molasses in the mixture.

Identification of Protein Sources

India is fortunate in having large amounts of crop residues high in protein, most of which have a fair degree of protection brought about by the processing methods. These materials are convenient to use in existing feed mills with the available equipment (i.e. grinders, mixers and pelleters) and the pelleted supplement is readily accepted because of a well developed marketing strategy.

In many other countries, particularly in extensive grasslands or savannahs, where major constraints to production of cattle are essentially the same as those for cattle fed crop residues, protein sources may not be readily available, or the sources not so obvious or easily obtainable. Most legume forages, legume seeds, edible tree leaves, seed pods and seeds that are available in these areas, contain highly soluble proteins which is easily fermented in the rumen. These, when used as supplements, only provide ammonia and minerals (they have for example 0.5-1% phosphorus). Fed without processing, because of their influence in the rumen they generally increase production of cattle on a basal diet of low protein roughage, but they provide little bypass protein unless they are fed as a large proportion of the total diet. Research results in Colombia fit into the general concept where supplementation of cattle on green *Brachiaria decumbens* pasture with either urea/molasses or the foliage of the fodder tree *Glyricidia*, increased production to the same extent.

Table 2: The effects of feed supplementation on livestock gain of cattle (6 per group) grazing on green *Brachiaria decumbens* pastures in the wet season (with mineral supplements) with liquid molasses/urea 10% or *Glyricidia* foliage.

Treatment	Rumen Ammonia (gN/1)	Initial Wt (kg)	Final WT (kg)	LWt gain (g/d)
No supplement	50	194	244	580
+ *Glyricidia*	170	204	266	717
+ Molasses/Urea	250	203	269	751

(Source: ICA, 1988)

Protein that is fermented in the rumen yields 80–100 g of microbial protein/kg of protein consumed. Whereas carbohydrate yields 180–200 g of microbial protein/kg of carbohydrate. On a high nitrogen diet, a significant supplement of soluble protein could in fact imbalance the P/E ratio because of the low microbial growth efficiency. This is a possible explanation for recent results reported from Iceland where fish silage and fish meal were compared as protein supplements to 'high quality' grass silage fed to young cattle. Fish silage, with the same N content as a fish meal supplement, depressed liveweight gain by 90g/day whereas a fish meal supplement improved liveweight gain by 360 g/day at the same N intake.

Table 3: The effect of supplementation of a basal grass silage * diet with fish meal or fish silage on liveweight gain and efficiency of feed conversion (after Olafsson and Gudmundsson, 1990).

Suppleent	Intake (kg/d) Supplement		Silage DM	LWt gain (kg/d)	Calculated ME intake	Efficiency (gLWt gain/MJ ME)
	DM	CP				
	.200	.123	4.6	.838	50	17
Fish	.400	.246	4.5	.914	51	18
silage	.400	.138	3.4	.358	40	9

* Early cut, precision chopped and preserved with formic acid, DM In vitro DMD. = 73% Calculated M/D = 10.4. CP = 16.8%. Cattle were Galloway/Icelandic cross - 6 months (8/group treatment), 160–205 kg LWt at beginning of experiment.

The above discussion defines and highlights two potential strategies to provide the two types of supplement required to optimise the efficiency of feed utilisation of cattle in areas with scarce resources of nitrogen or protein. The strategies must be to find forages/trees/tree seeds or pods that are high in protein and minerals. To then use this material in small supplements to provide for either the rumen with soluble protein and minerals or bypass protein after treatment to protect the protein. These uses can be combined with either a molasses/urea block and/or a source of locally available bypass protein.

Processing of Local Protein Resources to Provide Bypass Protein

The protein and render them non-fermentable in the rumen but allow them to retain digestibility in the intestines. In general, these fall into chemical treatment with agents that cross link with amino acids on the protein chain and include formaldehydes and aldehydes, tannins and simple sugars such as glucose and xylose. These reactions often require the protein source to be heated during processing.

Heat alone will often denature the protein to effect protection. For instance, Goering and Waldo (1974) found significant effects of the temperature used to dry lucerne on the subsequent animal production from that lucerne. In general, the higher the temperature of drying the greater the retention of protein by the animal. More recently, Lewis *et al.* (1988) have demonstrated that mild heat with a small amount of xylose is very effective in protecting soya bean meal protein. Xylose can be readily produced by acid hydrolysis of many fibrous materials including bagasse and cottonseed hulls.

Table 4: Effects on liveweight gain of cattle supplementing a basal forage/ concentrate based diet with soyabean meal or soyabean meal treated with sulfite liquor (SL) at 200°F for 2 hours (Lewis et al 1988).

	LWt gain(g/d)
No supplement	591
+7% soyabean	673
+9% soyabean + 10% SL	823
+8% soyabean + 5% SL	841

Little work has been done with forage proteins, although there are indications that the presence of low levels of tannins in the forage (e.g. leucaena) does afford some protection of its protein.

There is a great need to research methods for protecting the protein of forages, tree leaves, seeds and pods from local resources, as these are the proteins that are potentially available in the pasture areas of the world.

Trees as a Source of Protein and Soluble Nitrogen in the Rangelands

A number of authors have pointed to the small percentage of the total pasture biomass that is actually consumed by grazing animals in the extensive arid and semi-arid rangelands (Ellis and Swift, 1988). Most tropical grasslands are highly leached and their pastures apparently have low potential to provide ruminants with their required nutrients.

The point to be stressed is that 10–30% of pasture biomass is often all that is used by grazing animals. Trees, however, can produce considerable amounts of edible biomass. For example, the tree *Prosopis juliflora* produces up to 440 kg of edible pods per annum with an average 16% crude protein, (Riveros pers. comm) but the protein is, in all probability, soluble. Compare this with the usable biomass of 250 kg DM from the poor native pastures of South America and the combination of trees and grassland would obviously be a desirable development and synergistic for cattle production.

The research needs are obvious. Local research in centres of cattle density must identify actual or potential protein sources; must then establish mechanisms for harvesting, processing to concentrate the protein if necessary and protect it from rumen degradation. Appropriate means for using both the processed and unprocessed protein/minerals to optimise the efficiency of animal production must be established.

Practical Technologies for the Optimal use of Tropical Pastures and Rangelands

Food production in some tropical regions has risen dramatically during the last two decades as a result of the green revolution, which consisted of intensified production with increased inputs, such as, fertilisers, improved cultivars and crop protection. The tropical pasture revolution, hailed in the 1960's, has also resulted in an increased potential for producing livestock products. However, production improvements have been less obvious than in the crop sector. This can be attributed to many factors, notably:

- there is both a greater need and market for staple foods, such as rice, and cash crops than for food of animal origin,
- livestock production is usually carried out in drier areas and on poorer soils, and
- in many countries animal products are regarded as luxury rather than staple food items.

Chapter 7

Available Technology and its Application

Tropical pasture technology aims at improving animal production through the judicious use of inputs according to the prevailing economic and social conditions effecting the production system.

Forage production can be improved by establishing a) grass pastures requiring fertilization with NPK, b) grass-legume mixtures, c) or oversowing legumes into existing pastures, d) monocultures of legumes (protein banks) and e) fodder crops. The use of fertilisers on legume based pastures would consist primarily of Phosphorus (P) and rarely Potassium (K) plus trace elements. For intensive production systems on poor soils, fertilization would materially increase both the yield and persistence of pastures and would provide a good return on investments. Grasses are, however, more difficult to establish and are less persistent without fertiliser (particularly N) than legumes, although some legumes will not persist on poor soils without P and trace element supplementation. However, in many situations fertilization is not practical because of the low return from animal products, in such cases legumes such as *Stylosanthes* species can introduced which will produce even under low fertility. How the forage will be utilised (grazing or cut-and-carry) is also important. With grazing nutrients are returned, albeit very unevenly, but this is rarely the case with cut-and-carry systems where the nutrients are removed along with the forage and dung (and urine) is used for other purposes (fuel or manure). Repeatedly cutting forages rapidly depletes soil fertility and fertilization is necessary to sustain production.

Forage production is not dependent on pastures and forage crops alone and crop residues can make an important contribution in integrated crop/forage production systems. Two examples of the latter are the integration of livestock under tree crops (Reynolds, 1988) and alley farming.

The most appropriate method of pasture improvement depends on the intensity and the objectives of the production system. Dairy production for the liquid milk market, with assured income and high requirements for quality feed, can justify more intensive production methods than say beef production in remote regions which rely on uncertain markets for low quality beef.

The dairy farmer is more likely to base production on intensive, fertilized grasses or grass/legume mixtures; whereas the beef producer might consider, if anything, oversowing legumes into native pastures without fertilisation.

Numerous selected grass and legume varieties have become available since the 1960's for all tropical and subtropical regions with the exception of the semi-arid and arid zones (Henzell and 't Mannetje, 1980; 't Mannetje, 1984). Appendix 1 lists the main grasses and legumes for different climatic zones as defined by Troll (1966):

- tropical rainy climates with 12 to 9.5 humid months (Zone V_1)
- tropical humid-summer climates with 9.5 to 7 humid months (V_2)
- wet and dry tropical climates with 7 to 4.5 humid months (V_3)
- permanently humid climates with hot summers and maximum rainfall in the summer (IV_7)
- climates with 9 to 6 humid summer months and dry winters (IV_4)

The main requirements of forage species are that they are adapted to the prevailing climatic and soil conditions, tolerant of grazing, disease and pest resistant, free of harmful constituents (non-toxic) and tolerant of low fertility. It is not possible to find species to which all these requirements apply, however, management can make up for species' shortcomings.

For example, legumes which have a low grazing tolerance could be replaced by those that do posses it or grazed intermittently. Another example is the mimosine content of *Leucaena leucocephala*, which need not be a problem to animal health so long as the animal's diet does not exceed 30 % of the diet for more than six months. Alternatively, it is also possible to infuse the rumen of susceptible animals with micro-flora which can break down the toxic derivative of mimosine (DHP) (Jones and Metgarrity, 1986). Not all species need to be tolerant to low fertility because fertilizers can be used in certain production systems.

The Application of Pasture Technology

The decision to improve pastures in existing farm systems is governed by the following conditions:

- the need for improved animal production
- the value and marketability of animal products
- the motivation of farmers
- land tenure, and
- the availability of resources.

The Need for Improved Animal Production.

Animal proteins are an important component in the human diet and in most developing countries there is a need for both more and better quality food. Ruminants are able to produce food, especially proteins, from land which would otherwise be of little use for food production. When comparing the distribution of people, farm animals and the production of meat and milk between temperate (mostly developed) and tropical (mostly developing) countries, it is clear that tropical countries lag behind. Although far more animals are kept in the tropics, the production of meat and milk is only 36% and 18%, respectively, of the total world production.

This is due to inadequate nutrition, health care and management of the animals. Therefore, if animal production is to be increased, not only for meat and milk, but also for draught purposes and manure, then the feed supply must be improved. Since forages and crop residues are the primary feed sources, it is evident that both production and quality of forage needs to be increased.

An increasing demand for forages will also result from a) the increase in the demand for food as human populations continue to expand, and b) increasing purchasing power which leads to a greater propensity to buy animal products.

The need for pasture improvement can be shown by the low levels of animal production in both developed and underdeveloped tropical regions. This will be illustrated by examples from northern Australia and Africa.

Table 1: Distribution of farm animals, people, production of meat and milk (source FAO Yearbooks)

	% in developed countries	% in developing countries
Cattle	36	64
Buffaloes	—	100
Sheep	49	51
Goats	6	94
Pigs	40	60
People	25	75
Milk production	82	18
Meat production	64	36
Kg milk/person	320	23
Kg meat/person	54	11

Northern Australia: In northern Australia with its sub-tropical to tropical climate, large scale beef ranching is practised on unimproved native pastures. In the Northern Territory the average property size is 2500 km², the stocking rate is 6–8 beef animals per km² and the main emphasis is on breeding. Branding (weaning) percentages range between 40–60%, offtake ranges between 22–26% for animals aged between 3 to 4.5 years old and with dressed carcass weights between 225–360 kg. Most of the meat produced is exported as manufacturing beef ('t Mannetje, 1982).

By oversowing a legume into native pasture, or by establishing grass-legume pastures, substantial improvements in animal production can be obtained. Edye and Gillard (1985) reported that by oversowing *Stylosanthes hamata* cv. Verano into the native pasture along with the use of phosphate fertilizer, the carrying capacity could be increased up to ten-fold and cattle fattened in about half the time when compared with native pastures.

In an experiment at the Narayen Research Station in south-east Queensland three herds of breeding cows were used to measure beef production on unimproved native pasture at a stocking rate of 0.17 cows/ha and sown grass-legume pasture with superphosphate at stocking rates of 0.50 and 0.68 cows/ha over a period of ten years (Coates and 't Mannetje 1990). A summary of the results is given.

On both treatments conception and calving rates were high, but the long term average calving percentage of 70–75% for commercial beef production in the region on native pastures is also considerably higher than the Queensland state average (c. 60 %). All production parameters on sown pasture were higher than on native pasture. The results are most strikingly presented in terms of production per ha. Sown pasture produced 4.5 times more weight of calves per ha than native pasture and they were 20 % heavier.

Africa: In Africa most livestock are kept by smallholders for numerous reasons, the most important being milk and manure production, although traction and beef production may be locally important. Also throughout Africa cattle act as capital reserves and offer security of income or survival. Animal husbandry practices differ greatly between traditionally kept herds and commercial ranching, with corresponding differences in production, which can largely be attributed to disease control and animal husbandry. Birth weight and liveweight gains in indigenous cattle are low for both production systems, but particularly for traditional systems where the sale of animals for beef is not the primary objective.

Table 2: Mean conception and calving rates, average daily gain of calves, weaning weight (7 months of age), weight of calf per cow mated and weight of calf per ha on unimproved native pasture and sown grass-legume pasture in south-east Queensland.

	Native pasture	*Grass-legume pasture*
Conception rate	84 %	90 %
calving rate	79 %	82 %
Average daily gain calves	0.78 kg	0.92 kg
Weaning weight	200 kg	241 kg
Weight of calf/cow mated	155 kg	199 kg
Weight of calf/ha	26 kg	118 kg

The Value and Marketability of Animal Products

Nomadic herdsmen and small farmers in Africa are usually subsistence producers with limited access to markets. It is unlikely that these producers will have the necessary inputs required for pasture improvement.

Large scale producers are by definition dependent on a market for their produce. Price in the local, national and international markets will be determined by supply and demand. Local demand will, in turn, depend on consumers' incomes, therefore, in most developing countries producer prices for animal products tend to be low and there is little incentive to invest in pasture improvement.

Table 3: Calving rates, birth, weaning and live weights of different breeds and production systems in Africa (Brumby and Trail 1986).

Country	Breed/system	Calving %	Weight (kg)			4 yrs
			Birth	Weaning	2 yrs	
Mali	Sudanese/Fulani					
	traditional	54	17	55	125	200
	ranching	77	21	79	220	280
Nigeria	White Fulani/					
	traditional	46	20	55	140	240
	ranching	89	24	96	245	350
Ethiopia	Boran/					
	traditional	55	20	55	150	260
	ranching	78	25	180	265	420
Botswana	Tswana/					
	traditional	46	26	120	260	300
	ranching	74	31	180	360	400

The Motivation of Farmers

An important consideration of a producer contemplating pasture improvement is the cost and the expected return on investment. Costs are high and it depends on the financial situation of the farm whether it would be attractive to apply the technology. Although, eventually, a high return may be achieved, during the first few years after pasture improvement there will be a negative cash flow (Wicksteed, 1986). Not all farmers are in a position to carry this and credit facilities are not always available. There is also a risk involved that the improvement may fail, or that beef prices will decrease.

D.B. Coates (pers. comm.) listed as the main reasons for the present interest in pasture development based on legumes in Queensland (Gramshaw and Walker 1988) as:

- better beef prices providing money for investment;
- change in management brought about by compulsory testing for brucellosis and tuberculosis;
- high cost of mustering on large properties;

- market pressures for higher quality beef and faster turnoff rate;
- more emphasis on business aspects of the grazing industry;
- simple technology of oversowing *Stylosanthes* into native pasture at relatively low cost;
- new approaches to extension; and
- "successes" are increasing and so building up confidence.

Land Tenure

Land tenure can be a severe limiting factor to pasture improvement. If the producer does not have security of tenure then it is unlikely that he will be inclined to invest money into long-term pasture improvement. Most grazing lands in Africa are open to common grazing by privately owned herds.

This system works well with low human population pressures and therefore low animal numbers in a region. However, as human and animal population increase so does the pressure on the land and, since individuals are not responsible for the control of the grazing or for the maintenance of the land, everybody tries to maximise their return from the communal resource.

This leads to severe overgrazing, low productivity and danger of soil erosion. When land is owned or leased by the producer, there is the possibility of matching stocking rates with the carrying capacities in order to maintain and improve grazing and to replenish fertility. The chance of achieving a sustainable production system is less with communal than with privately owned grazing land.

The Availability of Resources and Knowledge

Before pasture improvement can be undertaken, there has to be an infrastructure for the distribution of inputs, such as, seed and fertilizers. In many regions there is no seed production or a seed trade.

Pasture improvement is further constrained by the lack of finance and of knowledge by the farmers and the extension service.

It is therefore necessary to promote the extension of pasture improvement practices, to encourage the development of seed production and to educate bankers on the possibilities of pasture improvement.

Opportunities for Forage Production Improvement in Different Livestock Systems

Forage Production Systems

Based on a classification of Perkins *et al.* (1986) for Indonesia and modified by 't Mannetje and Jones (1989) for south-east Asia; a number of distinct forage production systems can be distinguished on a global scale.

- Extensive permanent grasslands are unimproved native pastures (rangelands), which receive no inputs such as irrigation or fertilizer. On privately owned or controlled land, management consists of controlled grazing, burning and the control of trees and shrubs. Whereas communally grazing, which includes the savannas in Africa as well as road/canal sides and open forests, there are no forms of control and there is frequently overgrazing.

- Semi-intensive forage production systems have inputs, such as, fertilizer and weeding also irrigation (primarily applied to the main cash crop) which also benefits associated forage production.

- Intensive forage production systems have inputs applied for the sole purpose of forage production.

The ability to improve forage production is a function of climate, soils and the production system. Water and nutrients are physical prerequisites. Under natural conditions improved forage production is possible where rainfall is in excess of 650 mm/yr with a dry period not exceeding six months. In drier regions irrigation can be used. The existing production system, particularly the type of land tenure, also determines whether forage production can be improved and the type of inputs that can be realised.

Livestock Production Systems.

Livestock production takes place in three main systems:

- Pastoralism (nomadism, ranching, dairying)
- Livestock-crop systems
- Crop-livestock systems

Pastoralism: The main feature of this system is that the producers are entirely dependent on livestock for all their needs, there is no supplementary food crop production. There are three contrasting systems within pastoralism, *viz.* nomadism, ranching and commercial dairying.

Nomadism: is the way of life of indigenous peoples in arid and semi-arid regions who have no permanent place of settlement and move with all their livestock and possessions in search of water and forage. Regions in which nomadism is practised are characterised by an environment where the primary productivity which is so low that people cannot avail themselves of their feed requirements within a day's reach of a permanent settlement. The animals provide blood, milk, meat and income from the sale of surplus stock. The main source of animal feed is derived from extensive, unimproved, communally grazed grasslands. Post-harvest nomads are allowed to graze their animals on stubbles to consume crop residues and weeds and to deposit manure. However, nowadays, most crop farmers own livestock themselves and nomads are becoming less welcome. There are no real means to improve the feed base since the land is in communal resource.

Ranching: In Australia, Africa and South America ranching is undertaken in the sub-humid, semi-arid and arid regions to produce beef and wool. Properties are often large (1,000 ha to thousands of km^2), particularly in semi-arid and arid regions and may carry from 500 to several (20–50) thousand cattle or sheep and are usually either on long term leases or freehold. Production is extensive, with low inputs and corresponding low productivity both per animal and per unit area. The main feed supply are extensive unimproved native grasslands. However, ranching has the potential for sustained use of such grazing lands plus the opportunity to improve part of it by either establishing grass-legume pastures, by oversowing with a legume or by establishing a protein or fodder bank. The area for improved forage production will usually be small compared to the total area of the property and a judicious management system is necessary to optimise animal production and to increase the sustainability of the production system (Rickert and Winter, 1980; Coates and 't Mannetje, 1990). The extensive unimproved grassland can be used as the basic forage resource, with supplementation from small improved areas of grass-legume mixtures, protein banks or fodder crops for special classes of livestock (breeders, steers for fattening) at times of nutritional stress. Larger areas of legumeoversown grasslands can be used to either increase the overall carrying capacity or to reduce the area of land required for the existing herd (Edye and Gillard, 1985).

The reasons for improving forage production are to supplement the basic forage resource with more and/or better quality feed for various purposes, which include:

- to reduce mortalities which occur during dry times and where extra feed from improved pastures, protein banks or fodder crops can be used either for the whole herd to provide (near) maintenance rations or for a nucleus of animals to ensure continuation of the herd;

- to increase the conception and calving rates where an increasing plane of nutrition leading up to conception can increase fertility and reduce calving intervals;

- to keep the herd on a smaller area of land during the whole year, or part of the year (by oversowing part of the area) in order to reduce mustering costs; and

- to fatten younger stock on improved pasture or on feed from fodder crops or protein banks.

Table 4: Forage production systems used in different livestock production systems.

	LIVESTOCK SYSTEM			NRDLC	
	O	A	A	I	R
	M	N	I	V	O
	A	C	R	.	P
	D	H	Y	C	-
	I	I	I	R	L
	S	N	N	O	I
FORAGE PRODUCTION SYSTEM	M	G	G	P	V
EXTENSIVE PERMANENT GRASSLAND					
privately used	-	+	+	-	-
communally used	+	-	-	+	+
SEMI-INTENSIVE PERMANENT FORAGES					
understorey tree crops	-	-	+	-	+
forage from shade trees	-	-	+	-	+
forage in alley cropping	-	-	+	-	+
edges of crop fields	+	-	+	+	+
SEMI-INTENSIVE ANNUAL FORAGES					
after harvest crop	-	-	-	+	+
crop residues	+	-	+	+	+
INTENSIVE PERMANENT FORAGES					
improved grasslands	-	+	+	-	-
protein banks	-	+	+	+	+
INTENSIVE SHORT-TERM FORAGES					
fodder crops	-	+	+	+	+

Commercial milk production: is practised in regions of better rainfall and with access to markets, mostly near to urban centres. In humid regions dairy farms use improved pastures consisting of tropical species whilst in subtropical regions winter pastures consisting of irrigated temperate species may be appropriate (Stillman *et al.*, 1984). On tropical pastures milk production never exceeds 15 kg milk/cow/ day without the use of concentrates and usually it is less than 10 kg; whilst on unimproved grasslands production is very low (3–5 kg milk/ cow/day) (Stobbs and Thompson, 1975).

Livestock-crop systems: In regions with better rainfall, instead of nomadism people obtain their livelihood from a combination of livestock rearing and cropping. In these transhumance or seminomadic systems people have permanent settlements with some food cropping on better soils and a few animals for susistence, but with the main herd being moved to distant grazing lands in search of water and forage and returning to the settlements in the wet season. The main feed is obtained from extensive communally grazed grasslands yet there are opportunities to establish protein banks and fodder crops on privately owned crop land which also supply crop residues.

Crop-livestock Systems: These occur in regions with relatively high rainfall or a reliable wet season. The main activity is crop production although there may also be a high domestic livestock population. In the humid tropics rice and, in drier regions, other cereals such as millet, sorghum or maize are the main food crops. Because of high human population densities there is often a land shortage and farms are usually small therefore crop production is intensive. Animals (cattle and buffaloes) are needed for draught power, extra income, food and dung which may be used as either a fertiliser or fuel. The forage for the animals is obtained from waste land, road sides or canal banks. In some countries (e.g. Malaysia) there are publicly owned grazing reserves or areas of abandoned crop land, often covered in weed grasses such as *Imperata cylindrica*, are used as grazing lands (as in The Philippines and Indonesia). In west Africa grazing land is often fallow crop land.

Tree crop plantations offer great opportunities for the production of forage, particularly under mature coconuts and newly planted oilpalm before the canopy is completely closed (Reynolds, 1988). In Indonesia coffee plants are shaded by *Leucaena leucocephala* and pruned branches provide fuelwood and leaves for animal feed. Similarly, alley farming also offers opportunities for livestock feed production (Sumberg and Atta-Krah, 1988). Crop residues are important as forage and crop by-products are available as concentrate feed.

There are few other opportunities for improving forage production because of the lack of land, however, leguminous trees and shrubs can be used as living fences or as single trees in back yards. Depending on the need for land for food or cash crops and the price of livestock products, it may be profitable to grow fodder crops.

Table 4 shows which forage production systems can be used in the different livestock systems. This indicates clearly that commercial dairying and crop-livestock systems have the greatest range of forage production systems at their disposal and that nomadism has the least options.

In each production system there is a basic resource which provides the bulk of the forage. In nomadic and ranching systems, as well as in livestock-crop systems, the unimproved extensive grasslands, which are too poor for cropping, provide the basic forage although yields are low. There is no point in fertilising native vegetation since the response will not repay the cost of the fertiliser (Henzell and 't Mannetje, 1980). On the other hand, there is little export of nutrients from extensively grazed pastures and management must aim at sustainability through controlled grazing, burning and control of woody weeds.

Extensive grasslands are, however, the last major land resource in many regions of the world and with increasing population pressure these grasslands are being converted into marginal crop lands - such areas are liable to nutrient exhaustion and erosion. Controlled land use will remain an impossibility whilst there are communal rights of access because it is not in any individual's interest to reduce his use on the land if others continue or even increased their usage. The only solutions are controlled use and management of the grazing lands either instigated by the community using it or by government and population growth control.

Low cost forage improvement technology based on legumes with minimal fertiliser input is available. However, in developing countries serious constraints include communal land use, the low value of animal products and the lack of knowledge and resources, including finance. Integrated systems of crop and forage production and integrated use of forage production systems need to be further developed. The extensive unimproved grasslands (rangelands) are a precious resource which is under pressure of overuse and conversion to crop land.

Table 5: Selected grasses and legumes for pasture improvement in four climatic regions as defined by Troll (1966) ('t Mannetje and Jones 1989)

Grasses	*Climatic Zones*			
	V_1 & V_2	V_3	IV_7	IV_4
Andropogon gayanus	x	x	-	-
Brachiaria brizantha	x	x	-	-
Brachiaria decumbens	x	x	-	-
Brachiaria dictyoneura	x	x	-	-
Brachiaria humidicola	x	x	-	-
Brachiaria mutica	x	x	-	-
Brachiaria ruziziensis	x	x	-	-
Cenchrus ciliaris	-	x	-	x
Chloris gayana	-	x	x	x
Cynodon dactylon	x	x	x	-
Cynodon nlemfuensis	x	x	-	-
Digitaria decumbens	x	x	x	-
Digitaria setivalva	x	x	-	-
Panicum maximum	x	x	-	-
Panicum maximum var. trichoglume	-	-	x	x
Paspalum dilatatum	-	-	x	-
Paspalum plicatulum	x	x	x	-
Pennisetum clandestinum	-	-	x	-
Pennisetum purpureum	x	x	-	-
Setaria sphacelata	x	x	x	-
Sorghum almum	-	-	-	x
Sorghum sudanense	x	x	x	x
Tripsacum laxum	x	-	-	-
Urochloa mosambicensis	-	x	-	-
Zea mays	x	x	x	-
Shrub Legumes				
Albizia lebbeck	-	x	-	x
Calliandra calothyrsus	x	-	-	-

Codariocalyx gyroides	-	x	x	-
Flemingia macrophylla	x	-	-	-
Gliricidia sepium	x	-	-	-
Leucaena leucocephala	x	x	x	x
Sesbania grandiflora	x	-	-	-
Sesbania sesban	x	-	-	

Herbbaceous Legumes

Aeschynomene americana	x	-	x	-
Aeschynomene falcata	-	-	x	x
Alysicarpus vaginalis	x	x	-	-
Arachis pintoi	x	x	-	-
Calopogonium mucunoides	x	x	-	-
Cassia rotundifolia	-	x	-	x
Centrosema acutifolia	x	x	-	-
Centrosema macrocarpum	x	x	-	-
Centrosema pascuorum	-	x	-	-
Centrosema pubescens	x	x	-	-
Clitoria ternatea	-	x	-	-
Desmodium heterocarpum	x	x	x	-
Desmodium heterophyllum	x	x	-	-
Desmodium intortum	-	-	x	-
Desmodium ovalifolium	x	x	x	-
Desmodium triflorum	x	-	-	-
Desmodium uncinatum	-	-	x	-
Lablab purpureus	x	x	x	x
Lotononis bainesii	-	-	x	-
Macroptilium atropurpureum	-	x	x	x
Macroptilium lathyroides	-	x	x	x
Macrotylome axillare	-	x	x	x
Medicago sativa	-	-	x	x
Mimosa pudica	x	-	-	-
Neonotonia wightii	-	-	x	-
Stylosanthes capitata	x	x	-	-
Stylosanthes hamata cv. Verano	-	x	-	x

Stylosanthes humulis	-	x	-	x
Stylosanthes guianensis	x	x	x	x
Stylosanthes macrocephala	x	-	-	-
Stylosanthes scabra	-	x	-	x
Trifolium repens	-	-	x	-
Trifolium semipilosum	-	-	x	-
Vigna parkeri	-	-	x	-
Vigna unguiculata	x	x	-	x

Practical Technologies for Mixed Small Farm Systems in Developing Countries

In animal production systems, sustainable agricultural practices are those that promote systems in which there is efficient use of the natural resources based on an understanding of the interrelationships within prevailing agro-ecosystems. These need to be identified with agricultural and rural development processes that have an integrated approach to natural resource management in which potentially important technologies are technically appropriate, economically viable and socially acceptable to target beneficiaries, the small farmers.

In small farm systems, the objective of achieving sustainable animal production involving crops and animals assumes two major considerations. Firstly, it is essential that the available animal genetic resources within the mixed small farm systems will be fully exploited in a manner that is consistent with their biological attributes and potential productivity. Secondly, the choice of practical technologies should be realistic of the needs of such systems in order to promote sustainable productivity, as well as provide economic stability.

The prevailing circumstances suggest, however, that these factors are far from being in place and is reflected by the a) inefficient use of ruminants (buffaloes, cattle, goats and sheep), b) few examples of systems that are demonstrably sustainable, and c) concurrent low animal productivity. The problems are complex especially in small farms. These farms are limited by both access and availability to adequate resources. The situation is further exacerbated by the growth of both human and animal populations, rural poverty and environmental degradation; all of which emphasise that the choice of practical technologies are especially important and can considerably influence the sustainability of the system.

Another dimension concerning sustainability is the livelihood of the farmers themselves. In extreme situations where land sizes are particularly small or where landless labourers or tenants are involved, ownership of animals provides the only means of sustaining their livelihood. Nomads, pastoralists and transhumant peasants fall into this category. In all regions, but notably in north Africa, Middle East and north Pakistan, India and China, to whom ownership of animals is the key to livelihood and survival. In these situation, sustainable practical technologies for animals assumes even greater significance in socioeconomic terms.

It is appropriate therefore to examine the role of animal production in small farm systems and, in particular, to identify practical technologies that can significantly contribute to more sustainable systems and environmental integrity. The choice of technologies needs to take cognizance of the importance of crop-animal interactions where both components are complimentary and beneficial in terms of productivity from the land.

The purpose of this paper is to focus on those technologies which have been used in various developing countries and which show potential value. Specific case studies are cited in Asia, Africa and Latin America which highlight the relevance of these technologies in small farm systems involving crops and animals.

Size of Holding

Farm size is an important consideration and actual size varies between and with regions and countries. In Africa, North and Central America and Asia, the highest percentage of holdings are less than one hectare in extent. In Asia and Asia the majority of holdings are less than 5 ha.

The smallest farm size occurs in Bangladesh, while households cultivating paddy in Sri Lanka have average holdings of 0.3 ha of land.

Associated with the size of small farms, is the marked imbalance between the total ruminant livestock units and the available permanent pastures between regions. In Asia including centrally planned economies, the situation is especially critical where livestock densities are the highest in the world (FAO, 1986).

These small farms constitute the backbone of traditional agriculture throughout the developing countries (Devendra, 1983).

Small size presents a serious constraint, partly because of lack of access to resources, but more particularly, because of the difficulties of promoting the application of practical technologies capable of a discernable impact.

Thus, the small size of the farms determine, to a large extent, by the scale and extent to which new technologies can be adopted, and this aspect needs to be considered in the formulation and application of suitable technologies.

Practical Technologies

Notwithstanding holding size as a constraint to production and sustainability, there exist a number of practical technologies that have been tried with varying degrees of success. Really good examples of proven and demonstrable technology are, however, few.

This is partly because they have been inadequately tested on-farm with farmers, but more particularly, because chosen technologies have not been considered in a holistic and systems context. Nevertheless, several potentially valuable technologies are apparent and these will be discussed in the following sections and can grouped into two broad production systems:

- situations involving arable cropping systems, and
- situations involving tree cropping systems.

Table 6: Distribution of Size of Holdings by Region in the Developing Countries (Fao, 1971)

Region (number of reporting countries is given in parenthesis)	Total number of holdings (10³)	% Distribution									
		Holdings without land*	Under 1 ha	1–2 ha	2–5 ha	5–10 ha	10–20 ha	20–50 ha	50–100 ha	100–200 ha	200+ ha
Africa (20)	13,110.9	1.6	39.8	27.8	22.5	5.5	1.7	1.0	0.1	0.0	0.0
N & C America (14)	6,008.0	2.6	18.6	8.8	9.9	6.8	7.5	14.0	13.1	9.7	0.9
South America (7)	8,383.6	0.2	15.5	12.5	20.5	13.7	12.7	12.6	5.3	3.2	3.7
Asia (16)	103,619.8	0.2	52.7	19.1	17.7	6.6	2.7	0.8	0.1	0.01	0.0

* Establishments with no agricultural land which raise livestock and livestock products

Table 7: Variation in the size of small farms in some countries in South East Asia (FAO/UNDP, 1976)

Country		Definition
Bangladesh	a)	Subsistence farmers-cum-sharecroppers < 0.4 ha.
	b)	Viable and potentially viable owners, 0.4 to 0.8 ha.
India	a)	Small farmers, 2 to 4 ha of dryland (1 ha wet = 0.8 ha of dryland) and whose annual income <Rs 2400[+].
	b)	Marginal farmers, 0.8 to 2 ha of dryland and annual income <Rs 1200.
	c)	Agricultural labourers, < 0.8 ha of dryland and annual income <Rs 1200.
Indonesia	a)	Average size was 1.7 to 2.0 ha.
Korea	a)	Less than 1 ha and income below 500,000 Won per annuM[++].
Nepal	a)	Terai, 4 bighas (2.5 ha).
	b)	Hills, 1.75 bighas (1.0 ha).
Philippines	a)	Average size ranges between 0.8 to 7 ha.
Sri Lanka	a)	Agricultural households, 1.2 ha of land.
	b)	Paddy cultivating households, 0.3 ha of land.
Thailand	a)	Non-canal-irrigated areas: 15 rai*. canal-irrigated areas: 10 rai and net cash income of 1000 Baht**.

[+] US$ = 8.50 Rs (approximately)

[++] US$ = 285 Won (approximately)

* 1 rai = 0.16 ha

** US$ = 26 Baht (approximately)

Situations Involving Arable Cropping Systems

The Three Strata Forage System

In dryland farming areas, a major constraint to higher productivity from ruminants is the unavailability of good quality feeds especially during the dry season and periods of drought. The development of feeding systems that can increase the supply of good quality forages are, therefore, especially important to improve the prevailing low level of animal performance.

This is exemplified by the situation in Bali which has approximately three million people and three rainfall zones. Twenty five per cent of the land area is semi-arid with a rainfall of 900–1500 mm/yr.

Farmers constitute about 70% of the total population and most of them practice mixed crop-animal farming. Among the ruminants, Bali cattle are particularly important and, in the dry parts of the island, income from livestock accounts for between 29–43% of total farm income. Farmers generally own 2–3 herd of cattle which are used for draught and beef production.

The feed resources for ruminants in Bali come mainly from native grasses, tree leaves and cereal straws. The dry matter productivity from these resources is generally very low (approximately 2–2.5 tonnes/ha/yr). Dry matter yields can be increased through the introduction of improved grasses as well as forage legumes, eg. leucaena (*Leucaena leucocephala*) and *Gliricidia sepium*.

Bali cattle generally feed on wayside grasses and crop residues. Live weight gains vary between 100– 200 g/day and marketable weights are only reached in about four to five years. However, with improved feeding and concentrate supplementation, it has been shown that daily weight gains of Bali cattle could be increased to between 400– 600 g/day and that the fattening period could be reduced to less than two years.

These circumstances led to the development and successful demonstration of the Three Strata Forage System, a project supported by the International Development Research Centre (IDRC) of Canada. The system involves a first stratum of grasses and ground legumes; a second stratum of shrub legumes; and a third stratum of fodder trees.

The project ran for five and a half years and several relevant results were found comparing two types of systems: The Three Strata Forage Systems (TSFS) and the non-TSFS (NTSFS) at two stocking rates (2 and 4 cattle/ha). The results of the project have been published (Nitis *et al.*, 1990; Lana *et al.*, 1990; Arga, 1990; and Nuraini, 1990). Some of the results, and the following summarises the main highlights:

- The allocation of 0.09 ha of land as a forage boundary within the TSFS produced 98% more forage than the 0.25 ha of natural pasture. With the additional 2000 shrubs and 42 trees, the total wet and dry seasons forage production in the TSFS was 91% more than that of NTSFS.

- *Stylosanthes, Centrosema, Acacia, Gliricidia and Leucaena* which were grown, provided increased dietary nutrients in the

TSFS. Consequently, cattle raised in TSFS gained 19% more live weight and reached market weight 13% faster.

- The availability of increased forages also enabled higher stocking rates and live weights to be achieved: 3.2 animal units (375 kg)/ha/year in the TSFS compared to 2.1 animal units (122 kg)/ha/year.

- Cattle in the TSFS were less infested by endoparasites. This was presumably due to less contact with the traditional cattle, since the TSFS cattle were always kept in confinement.

- The introduction of forage legumes into the TSFS reduced soil erosion by as much as 57% in the TSFS compared to NTSFS. Additionally, the fertility of the soil in the TSFS was also considerably improved.

- With the presence of 2000 shrubs and 42 trees lopped twice a year, firewood production from 0.25 ha of TSFS was 1.5 tonnes per year and TSFS supplied 64% of the farmer's requirement.

- Farmers in the TSFS spent less time managing their cattle and, more importantly, these same farmers benefitted by an increased 31% income compared to NTSFS farmers.

Further research and development aimed at creating a more sustainable TSFS will involve the introduction of goats. The justification for including goats is a) the official promotion of goats in dryland farming areas, and b) the potential for generating additional income. Additionally, goats will provide greater flexibility of resource use by farmers. On a live weight basis a 375 kg Bali bull is equivalent to 6 30 kg goats. When feed is limited during the dry season, it is easier to reduce the number of goats rather than cattle. Theoretical calculations suggest that with shrubs and tree fodders, increased carrying capacity is feasible and detailed studies are continuing on the utilisation of *Gliricidia* as a forage.

Table 8: Comparative Productivity of TSTS and NTSFS Plots (kg dry weight/ plot per year) (Nitis *et al.*, 1990)

Parameter	TSFS[+]	NTSFS[++]
Food	853	1268
Straw	750	1218
First stratum	455	-

Second stratum	310	-
Third stratum	15	-
Shrubs	-	132
Trees	-	2
Improved grasses	-	10
Native grasses	-	242
Firewood	1049	475
Cattle live weight gain (kg/3 years)	186	166
Carrying capacity (cattle/ha)	4	2
Maximum live weight (kg/head)	300	200
Soil erosion (mm/2 years)	11	20

+ Three strata forage system
++ Non-three strata forage system

Food-Feed Intercropping

The concept of food-feed intercropping in both lowland and upland small farm systems is relatively new. The two principal advantages are: a) that the system aims to provide sustainability through involving the complimentary role of crops and animals; and b) the use of appropriate forage crops provides fodders and crop residues which are valuable both ruminants and non-ruminants.

Already several attempts to develop food-feed systems have been undertaken notably in the Philippines and Thailand. Rice is the principal cereal crop, but other crops which have been used with the aim of increasing the production of animal feeds include: cowpea, maize, groundnut, pigeon pea, sorghum and sweet potato. The criteria for the choice of the inter-crop include *inter alia*: the type of animals reared, potential forage or crop residue yield, promotion of soil fertility, drought tolerance and extent of the dry season, shade tolerance in upland areas, ease of eradication and resource requirements. The strategy is to integrate within the rice cropping pattern (intercropping and relay cropping) other feed producing crops and forage crops without reducing the area of the land used. Exemplifies this situation in the Philippines.

In the past, the general tendency has been to grow only one crop of rice in both the rainfed lowland and upland areas resulting in a low cropping intensity. Both areas are important for livestock production. Furthermore, the rainfed lowland rice areas occupy about

67% of the total land area in Asia and where the bulk of swamp buffaloes, cattle and sheep are found.

The drier upland areas generally favour the presence of small ruminants, mainly goats, but also some large ruminants.

Rice and wheat straws are the main residues from cereal cultivation and these supply the basic diet for ruminants. They provide bulk and energy for maintenance but not production.

The intake of these roughages is limited by their low crude protein content (4–6%) and low digestibility, which necessitates supplementation to meet production requirements. Additionally, the reduced feed availability during the dry season and associated weight loss and poor performance, necessitates the application of strategies to ameliorate the situation and increase available dietary nutrients.

In the arable areas, cereal straws (mainly rice and wheat) and other crop residues are the most available and cheapest feeds for ruminants.

The greatest challenge rests with the development of effective and economic feeding systems which utilise the high lignocellulosic content of straws.

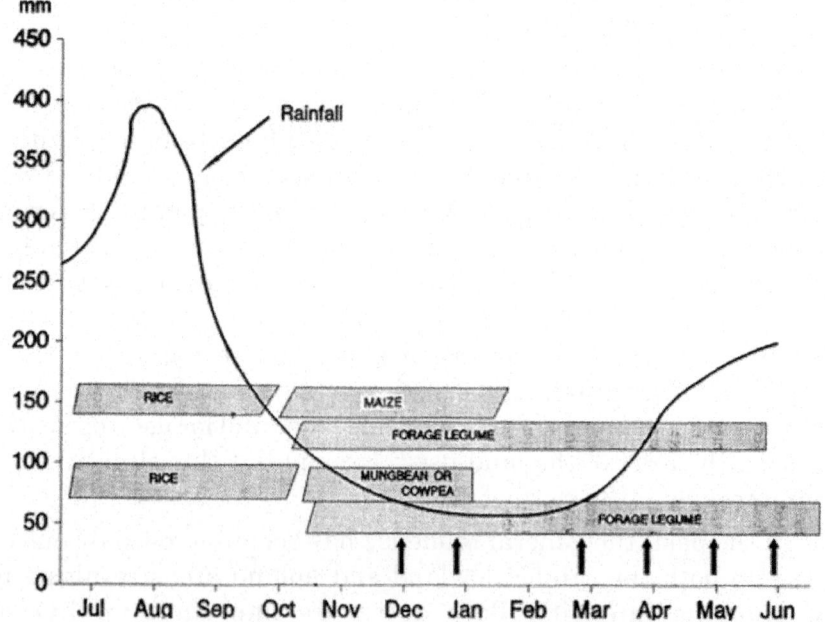

Figure 1: Cropping patterns involving rice and food crop - forage intercropping

The value of integrating two forages lablab (*Lablab purpureus*) and *Clitoria ternatea*.

The latter is grown by farmers in northern Philippines after rainfed lowland rice as monocrop for green pods and forage. No differences were found in mungbean grain and residue yield.

The combination of mungbean plus lablab (cv Rongai) gave the highest forage yield of 7.9 tonnes/ha.

Similarly, the results of an experiment on maize-forage intercropping under rainfed upland conditions (Tengco and Carangal, 1987) involving varying plant densities.

Four forage legumes were used: *Stylosanthes guianensis* CIAT 136, *Macroptilium atropurpureum* cv Siratro, *L. purpureus* cv Highworth and *Desmanthus virgaties*.

Of these, Siratro gave the highest forage yield.

The results demonstrate the value of intercropping with suitable forages to produce valuable feeds for animals.

Table 9: Grain, Residues and Forage Yields of Mungbean and Forage Legumes as Monocrop and Intercrop Combinations in the Philippines (tonnes/ha) (carangal *et al.*, 1988)

Crop Combination	Grain	Residues (DM)	Forage Yield (DM)					Total (DM) forage & residue
			Initial cut	Regrowth			Total	
				1	2	3		
Mungbean + lablab (cv. Hatiya)	1.02	0.91	1.69	1.98	1.11	0.91	5.69	6.60
Mungbean + lablab (cv. Rongai)	1.00	0.94	1.99	2.43	1.83	1.68	7.93	8.87
Mungbean + clitoria	1.28	1.14	-	1.53	1.16	0.93	3.62	4.76
Mungbean	1.26	1.19	-	-	-	-	-	1.19
Lablab (cv. Hatiya)	-	-	2.70	1.73	1.16	1.40	6.99	6.99
Clitoria	-	-	2.43	2.70	2.06	1.73	8.92	8.92
CV (%)[+]	18	13	20	32	20	32	10	
LSD[++]	NS	NS	0.89	1.17	0.49	0.71	1.79	

[+] Coefficient of variation

[++] Least significant difference (P < 0.05)

Table 10: Yields of Maize and Forage Legume Intercropping at Three Plant Densities during the Dry Season in the Uplands, Philippines (tonnes dm/ha) Tengco and Carangal, 1987)

Intercropping	Main crop		Intercrop
	Grain	Fodder	Forage
Maize 26,666 pl/ha monocrop	2.95	3.72	-
+ Stylo CIAT 136	3.14	3.91	1.84
+ Siratro	2.57	2.57	2.95
+ *L. purpureus*	3.19	3.02	1.54
+ *D. virgatus*	2.69	3.21	0.64
Maize 35,555 pl/ha monocrop	2.94	3.62	-
+ Stylo CIAT 136	2.27	3.36	1.14
+ Siratro	2.41	3.77	2.88
+ *L. purpureus*	2.19	3.28	1.14
+ *D. virgatus*	2.22	3.15	0.62
Maize 53,333 pl/ha monocrop	4.40	4.94	-
+ Stylo CIAT 136	4.23	5.09	0.60
+ Sirato	3.40	4.49	2.72
+ *L. purpureus*	3.70	3.89	1.69
+ *D. virgatus*	3.75	5.17	0.75
C.V. (%)±	28.4	23.4	47.2
F-test	**	**	**

[+] Coefficient of variation

[++] Level of significance

Relay Cropping

Relay cropping is an important means to increase the supply of feed for farm animals. In relay cropping, a second crop is planted into the first before harvest eg. the introduction of legumes (groundnut or pigeon pea) into a main crop of rice or wheat. The strategy can extend the supply of feed, possibly, throughout the year.

Integrated Pig-Ducks-Fish-Vegetable Systems

A sustainable system which is widely practised in South East Asia and China involves the integration of pig production with fish farming, duck keeping and vegetable production, or a combination of these (Devendra and Fuller, 1979). The inter-relationships between the components. The system is based on the use of ponds which not only meet the needs of pigs, but also enables fish and ducks to be kept. Water is also useful for vegetable production.

Such systems are especially suited for small farms where only a few pigs are reared together with either ducks or fish and intensive vegetable production. Edwards (1983) estimated that 26.7 laying ducks or 8.2 pigs or 0.8 dairy cow or 1.7 buffaloes are required to produce a mean yield of 174.7 kg of fish/year in a 200 m²pond, based on equivalent nitrogen manure inputs.

Situation Involving Tree Cropping Systems

Integrated Animal and Tree Cropping Systems

Integrating ruminants with tree cropping systems, such as, coconuts, oil palm and rubber are important production systems which have not been adequately exploited. Livestock/tree cropping systems are common in the humid and subhumid regions where intensive tree-crop production is practised. Although such systems are not new greater attention is needed to ensure more complete utilisation of the land. The advantages of the system are:

- increased fertility through recycling of nutrients (dung and urine)
- weed and shrub control
- reduced use of herbicides
- reduced fertiliser wastage
- provision of shade
- easier management of the plantation, and
- distinct possibilities of increased crop yields and animal products from the land.

There is an estimated area of 20.3×10^6 ha under tree crops in South and Southeast Asia which reflects the potential for this kind of activity (FAO, 1986). Many of the Pacific island territories, notably Papua New Guinea, New Hebrides, Fiji, the Solomon Islands and Western Samoa have large areas under coconuts and a potential for integrating ruminant production.

Malaysia provides a specific example of the economic benefits of integrating ruminants with oil palm where an estate has allocated a portion of the plantation to the workers for grazing their animals. For the first two years (1980 and 1981) only cattle were owned and grazed, however, in 1982 and 1983 goats were also introduced. This was done because of their economic importance and capacity to supply both meat and milk in the estate.

A comparison between the grazed and non-grazed area involving both young and mature trees is valid in that it involved both the same area of 71–135 ha and, more particularly, the fact that both areas were on the same soil type. The effect grazing cattle and goats was an increase 2.15–5.16 tonnes fresh fruit bunches per hectare per year over four years.

When translated into the total hectare available for grazing and sale value per tonne of fresh fruit yield, the economic advantage is substantial.

The result in economic terms is similar to the findings in West Java of integrating sheep and goats under rubber. The presence of legume is of definite advantage and it has been calculated that the amount of nitrogen utilised by the animal and excreted in the faeces and urine increases with the presence of the legume cover.

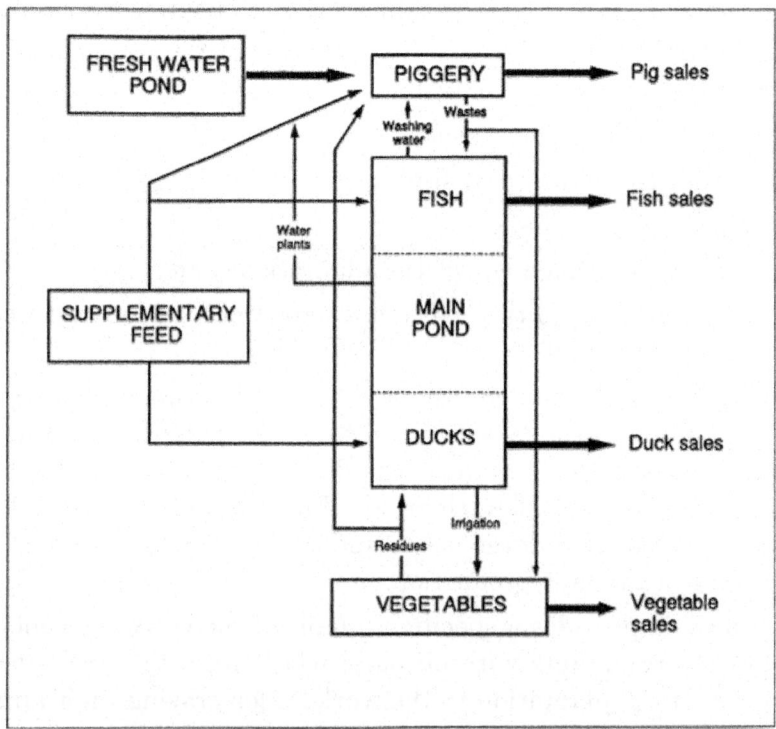

Figure 2: Integrated pig - fish - duck - vegetable system

Table 11: Effect of Mixed Cattle and Goat Grazing on the Yield of Fresh Fruit Bunches in Oilpalm Cultivation in Malaysia (Devendra, 1986)

Year	Yield of fresh fruit bunches (tonnes/ha)		
	Annual grazed area	Annual non-grazed area	Difference
1980	30.55 (cattle)	25.61	4.94
1981	17.69 (cattle)	15.87	1.82
1982	25.12 (cattle and goats)	22.97	2.15
1983	23.45 (cattle and goats)	18.29	5.16
Mean	24.20	20.69	3.51

More recently, a study has reported the integration of goats with pine (*Pinus insularis*) in the Philippines (Penafiel and Veracion, 1987). A stocking rate of four goats/ha did not cause significant soil loss nor disturbance of soil bulk density, compaction and infiltration rates.

Thinning of the pine stands increased tree growth as well as forage production. The daily weight gains were impressive and averaged 99 to 129 g/day/goat.

The integration demonstrated a ecological and economically viable and sustainable farming practice for the forest dwellers.

Alley Cropping

Alley cropping provides another opportunity to integrate crops and animals as well as enhancing sustainability. The development of the system has the following advantages:

• maintenance and promotion of soil fertility
• increase the supply of animal feeds
• supply of fuelwood, and
• contribute to the development of all year round feeding systems.

The results of a study in Nigeria involving leucaena (*L. leucocephala*), maize and grazing with goats and sheep.

The effects of both alley farming and alley farming after fallow on the yield of maize grains were 30–70% higher than that of conventionally cropped plots. The results did not separate out however, the specific effects of the presence of animals and fallow.

Table 12: Yield of Maize Grain in Conventional Farming, Alley Farming and Alley Farming after a Two-Year Grazed Fallow in Nigeria (tonnes/ha) (Reynolds and Attah-Krah, 1989)

Farming system	First season	Second season	Total
Conventional (no trees)	2.13	0.93	3.06
Continuous alley farming[+]	2.41	1.70	4.11
Alley farming after fallow	3.30	2.04	5.34

[+] Involving *L. leucocephala* and grazing by goats and sheep

Supplementation

In many developing countries, regular feed shortages and droughts are common. In such conditions, subsistence feeding mainly on cereal straws results in reduced live weight and perpetual animal low productivity. Inadequate nutrition is also associated with delayed age at first parturition, increased parturition intervals, prolonged non-productive life and high mortality.

Strategic supplementation of energy, proteins and minerals offer an important means to ensure that animal performance is not reduced, especially during critical periods of feed shortages. Several alternative strategies have recently been pursued, with the objective of better utilising low-quality forages for production (meat, milk or draught). Foremost in these initiatives are a variety of chemical pretreatments (Jackson, 1977; Ibrahim, 1983; Sundstol, 1984; Doyle and Pearce, 1985). Among these, urea treatment is currently the most widely used to improve nutritive value of straws. The result has been a shift from a subsistence to a maintenance levels of nutrition, achieved by increased feed intake and/or digestibility (Ibrahim *et al.*, 1984), and occasionally growth (Perdok *et al.*, 1984; Verma, 1983) and milk production (Davis, 1983) responses.

The use of forage supplements has been secondary to chemical pre-treatments and has been underestimated and not given adequate research and development attention. For a variety of reasons, this approach has enormous potential for ruminants, especially in situations where animals are abundant and varied. It is appropriate, therefore, to review current understanding of the use of forage supplements and the benefits of this strategy. There are many advantages concerning the use of these forages (Devendra, 1988) especially in mixed farms in the humid tropics:

- availability in the farms
- accessibility
- provision of variety in the diet
- source of dietary nitrogen, energy, minerals and vitamins
- laxative influence on the alimentary system
- reduction in the requirements for purchased concentrates, and
- reduced cost of feeding.

There are several good examples of such forages, especially legumes, that can be increasingly used in small farm systems. Leucaena, for example, provides a valuable source of protein, energy and sulphur for the rumen bacteria, as well as being used for fencing, fuel and mulch. Further important examples include: *Acacia* spp., *Ficus* spp., cassava (*Manihot esculenta* Crantz), *erythryna* (Erythryna spp.), gliricidia (*Gliricidia spp.*), leucaena (*L. leucocephala*), pigeon pea (*Cajanus cajan*) and Sesbania (*Sesbania spp.*).

The potential value of these forages has recently been reviewed by Devendra (1990) and a number of observations are worthy of mention:

- With all ruminants the use of various forage supplements consistently increased live weight gain or milk production,
- in many instances the beneficial response (meat and milk) was associated with a reduced production costs,
- of the forage supplements used, legumes have been particularly advantageous, and
- in the majority of situations stall-feeding or cut-and -carry systems are most common than grazing situation,

With regard to large ruminants, the research on the value of forage supplements for draught purposes is sparse. In Thailand, the effects of breed (Murrah X Swamp and Swamp) and feed supplement (with or without) on draught (work or no work) was studied over six months. The concentrate supplement consisted of cassava chips and dried leucaena leaf in a 3: 1 ratio and was applied at 1.5 kg/head per day. The animals without supplements were feed on silage. Over the first four months before the draught capacity was assessed supplementation, as one would expect, significantly stimulated growth rate in both breeds. Concerning working ability, Thai swamp buffaloes ploughed more than the crossbred Murrah, and supplementation increased the area ploughed. However, there was no significant breed effect. Neither were differences were found in the speed of ploughing between breeds or supplementation, although, there was a tendency towards faster ploughing for both breeds when feed supplements were used (Konanta *et al.*, 1986).

In terms of practical application, a recent review of the benefits of including high protein forages supplements to the diet suggest the following optimum dietary levels (Devendra, 1988):

-	optimum dietary level on DM basis:	30–50%
-	as % of live weight:	0.9–1.5%

Wong *et al*,

1987 clearly shows that Leucaena supplementation increases milk production and reduces feed costs.

All Year Round Feeding Systems

The strategy to increase feeds within small farm systems should have the final objective of developing sustainable all year round feeding systems appropriate to the prevailing situations. In this quest, maximising feed production is essential and the following approaches are feasible:

- intercropping with cereal crops
- relay cropping
- food-feed cropping systems
- intensive use of available crop residues
- forage production on rice bunds
- alley cropping, and
- three strata forage system in dryland areas.

The key elements in the use of these approaches are the quantification of the feeds produced throughout the year and efficiency with which they are utilised by the available animals. The former enables the determination of feed balance sheets, the extent to which animals can be supported and identify the critical periods during which feed deficits need to be corrected.

The scope for increasing the efficiency of feed utilisation in innovative systems is enormous. More innovative feeding practices are necessary that can sustain all year round feeding in more intensive systems of production. These could include the various chemical pre-treatments, supplementation and the use of multi-nutrient block (Kunju, 1986). Processing (chopping and grinding) of some crop residues may be feasible in certain situations and, associated with this, the development of complete feeds (Reddy, 1987). An overriding consideration that will determine the value of these approaches is the demonstration of economic benefits. For successful application and acceptance at the farm level, practical technologies need to be simple, within the limits of the farmers' capacity and resource availability, convincing and consistently reproducible.

The need for more evaluative work especially regarding the greater use of fibrous crop residues. Onfarm animal research (OFAR) is probably the only accurate assessment of whether new technology packages are acceptable both economically and socially to farmers, since the technique takes into account the interacting components within farming systems.

OFAR is also a means of identifying and addressing the constraints to adoption of new feeding systems and the extent to which they contribute to sustainability (Devendra, 1990b). Two phases are involved:

Phase 1: Information needed by farmers and advisers:

- Choice of feeds,
- extent of availability and seasonal supply,
- nutritive value
- detailed information on constituents
- prediction of change during production
- capacity to meet production target, and
- capability of filling any nutritional gap
- strategic supplementation:
- on animal performance, and
- on sustainable production of the basal feed resource
- feeding system,
- level and type of production, and
- realistic production targets.

Phase 2: Mechanisms for delivering information to farmers:

- Methodology, a balance between basic and applied research,

- linkages between disciplines and institutions,
- demonstration of economic benefit,
- large-scale on-farm testing,
- *in situ* utilisation
- farmer participation, and
- definition of a model for feed resource development.

Several criteria need to be considered regarding on-farm work and these include *inter alia* the following: production objectives, the treatments involved, the precise methodologies to be used, measurements to be undertaken, type and value of the inputs used and the output derived from the experiment, extent of farmer participation, issues related to the economic analysis of the results and marketing.

Chapter 8

Other Practical Technologies

In small farm systems, farmers can resort to several useful and practical technologies. This is a rational means to diversity the use of meagre resources and spread the risks. A number of these that have been proven successful are briefly mentioned in the following sections.

Soaking

Wetting or soaking cereal straws is a common practice and helps to increase feed intake. The addition to soaking the inclusion of 1% urea solution will further increase the nutritive value of the straw.

The technique has also been used to increase the utilisation of sorghum grains fed native pigs in El Salvador, where pig production is mainly a family enterprise and over 80% are indigenous breeds. Sorghum is soaked for 72 hours to allow fermentation and to increase digestibility, it is then fed with forage supplements (*Melanthera nivea*, *Ipomoea spp.* and *Desmodium spp.*). The results gave feed conversion efficiencies of 4.6, 5.0 and 4.2 for these treatments compared with 7.6 for the control group fed only unfermented sorghum (Ministry of Agriculture and Livestock, and Institute for Nutrition for Central America and Panama, 1986).

Sun Drying

Sun drying of some feeds before they are fed is a common practice among farmers in certain areas. The main reasons for this are to reduce the water content to ensure adequate dry matter intake. Good examples are the water hyacinth (*Eichornia crassipes*) and sweet potato vines (*Ipomoea batatas*) both of which are used extensively for feeding pigs in South East Asia.

The other main reason is specific to cassava leaves (*Manihot esculenta* Crantz) which contains hydrocyanic acid and is toxic to animals. Sun drying freshly harvested leaves and root peelings reduces hydrocyanic acid levels through the interaction with sunlight; the cyanogen is released as gas. A similarly, wilting leucaena leaves causes the enzymic hydrolysis of mimosine to 3.4 (1 H) pyridone, eliminatingacute but not chronic toxicity. A similar change occurs when the leaves are dipped in warm water (Lowry, 1990).

Feeding Mixed Forages

Farmers throughout the developing countries traditionally feed a mixture of forages, especially to small ruminants. The underlying reasons are associated with reducing any toxic effects found in any one feed (eg. cassava or leucaena leaves) and increasing the variety and palatability of the diet.

Water Conservation and Crop Production

In many parts of the developing countries, water is a critical constraint to crop production and its effect on animals is reflected in relatively low productivity in the semi-arid and arid regions.

In Mexico, three types of reservoirs to capture run-off water were tested in goat rearing semi-arid areas: ponds, micro dams and micro water sheds. Rainfall figures of 83.4,84.2 and 69 mm of water were recorded for the three types at 40 cm of depth respectively. The control rows only had 56.5 mm of water. The higher water retention allowed higher dry matter (DM) yields of maize (1744, 1794 and 1518 kg/ha, respectively) than the control (373 kg/ha). It was also found that the use of terraces allowed a 150% increase in the planting area. Maize was planted for human consumption and the residues were used for goat feeding. The biomass yield was 9.07 tonnes DM/ha, equivalent to 81 kg of DM produced per mm of rain.

The efficient conservation of water has also enabled the establishment of large scale trials in farmer's cooperatives where water capturing has assisted the growing of *Atriplex canescens* for feeding goats (INIFAP, 1989).

Sustainable animal production is dependent on the choice and application of appropriate practical technologies. The scale and extent of the success of the latter is influenced by the size of landholding. Present evidence suggests that good examples of practical technologies that promote sustainable systems are few, and more development effort is necessary to demonstrate their potential value and impact.

Such efforts need necessarily to consider the nature of the mixed farm operations, crop-animal interactions, viability of the system, efficiency in the use of meagre esources and opportunities for demonstrating impact.

Cost-Effective Disease Control Routines and Animal Health Management in Animal Agriculture

Developing countries, with a few exceptions, inherited livestock research and development systems they had little or no hand in designing. These have not, in the main, performed well.

Given current and probable future dependence on external donor support, programme conceptualisation in these countries is likely to largely remain in other hands. This is happening at a time when scientific manpower in the livestock sub-sector has been improving. At the same time, the better Third World scientists are faced with diminishing operational support and personal incomes and are opting out of public sector institutions and often out of their countries.

Rapidly increasing populations and socio-political instability in much of the developing world have added to the pressure to find lasting solutions.

Animal diseases were for many years recognised as important constraints to livestock production and formed a predominant portion of donor funded programmes. For the past two decades, increasing the numbers of veterinarians through the establishment of veterinary colleges and training facilities for paraveterinary personnel has been a major preoccupation.

In recent years however, views and estimates have been given about the number of animal health personnel needed to adequately service a given number of livestock units under various conditions (SEDES/FAO, 1977; Sandford, 1983). What the more recent studies state is that in absolute national terms, where the number of livestock units is simply divided by the number of veterinary personnel, many countries, even in Africa the worst endowed continent, are attaining the required theoretical ratios. Favourable national ratios on a macro-scale have not always been translated into expertise available at producer level. The most persistent of the reasons given for this are:

* Erosion of operation performance caused by too high a proportion of the budget being utilised on personnel costs.

- Policy issues which overburden the public sector with functions (and attendant expenses) which are better managed outside this sector (Leonard, 1984; Anteneh, 1985; de Haan and Nissen, 1985).

Table 1: Ratio of Livestock Units (×1000) to Government Veterinary Staff in Sub-saharan Africa.

Region	professional		para-professional	
	mid-1970s	mid-1980s	mid-1970s	mid-1980s
Western	95:1	30:1	7:1	4:1
E. & South	120:1	80:1	14:1	6:1

In research, which has over the years been much more dependent on external financing than field services and where the percentage of expatriates has been substantial, there have been major shifts in donor policy in recent years. After years of uncoordinated donor support recent moves have been towards promotion of regional cooperation and coordinated donor funding. Networking has been an important element in cooperative programmes. The CGIAR Centres have been promoted by most donors as the main hubs for networking activities.

Persistence of animal health problems in the presence of local experts has not boosted their morale or enhanced their standing in the eyes of the affected farming public. Frustration has ensured at various level and has tended to be especially pronounced among those who have greatest drive, superior ability or training. This has in turn encouraged staff instability due to movement out of public service to other sectors or reluctance to return home for those in training outside their countries.

Disease Control Routines

A primary reason for establishment of veterinary services in most of the developing countries was for the control of epizootic diseases. Rinderpest and contagious bovine pleuropneumonia (CBPP) were the first cattle diseases targeted.

As the animal health situation improved and livestock improvement programmes were instituted, other diseases such as foot and mouth disease became important. Another prerequisite for disease control was the establishment of diagnostic facilities. Still later, many countries took the decision to establish vaccine production laboratories.

Besides epizootic diseases, there was early recognition of the limits that a number of enzootic diseases placed on livestock improvement. These included many viral and bacterial diseases and arthropodborne protozoal/rickettsial diseases such as trypanosomiasis, theileriosis, babesiosis and anaplasmosis.

Through the existence of diagnostic facilities, many diseases hitherto unknown were identified. Not all were of national importance but which, in the absence of alternative disease control services, veterinary departments, being public institutions, felt obligated to deal with. The same rationale was subsequently used to extend public services to clinical veterinary practice.

Control of Epizootics

The Example of Rinderpest: Rinderpest is recognised as the most devastating of all livestock diseases and its control has served as a prototype for the control of other epizootics. Though now controlled in all continents with the exception of Africa and Asia, the potential devastating effect of this disease is well known. In the Great African Pandemic of 1880–95, rinderpest is estimated to have killed 80–90 percent of Africa's cattle and wild ruminants (Faulkner, 1983). The more recent epizootic of 1981–1983 is estimated to have caused an economic loss of US\$ 300 million (de Haan and Nissen, 1985). Given the devastating mortality rinderpest has caused over the years, justification for its control appeared self-evident not only for the particular country affected but for neighbours and importers of livestock and livestock products.

The method used for rinderpest control using large coordinated national and regional campaigns has been effective, although inability to reach all animals at risk has made it impossible to eradicate the disease.

Vaccination coverage is always maximal during actual or threatened disease outbreaks when the cooperation between livestock owners and vaccinators is at its peak. This cooperation does however fall off rapidly as soon as mortalities cease and cattle owners do not feel an immediate threat. The estimated vaccine coverage for rinderpest and contagious bovine pleuropneumonia (CBPP) in various African regions during the 1970s. One result of the decreasing vaccination coverage was that, by 1985, rinderpest had reoccurred throughout the Sahel, West, Central and East Africa.

Table 2: Rinderpest & CBPP Vaccination Coverage (%) in Sub-saharan Africa

Region/Country	Year	Rinderpest	CBPP
Sahel	1975–90	42	36
S/Humid Anglo.West Africa	1975–90	17	38
S/Humid Franco.West Africa	1975–90	2	15
Sudan	1970–80	20	na

Cost Effectiveness of Epizootic Control: As was demonstrated during the special rinderpest disease control programme of 1962–76 in Africa (Joint Project 15) and as is evident from results to date in the ongoing Pan African Rinderpest Campaign, that it is technically easy to control the disease, but control is much more cost effective in an atmosphere of political tranquillity. Attempting to reach areas of political conflict considerably adds to the cost and reduces vaccination efficiency.

There have been recent advances in rinderpest vaccine development aimed at further increasing the effectiveness and reducing the cost of rinderpest control. Among these is the recent demonstration of protection by a vectored recombinant vaccine (Yilma et al., 1988) and the reported thermostable vaccine whose economic value has recently been analysed (Stryker et al., 1988).

The likely benefits of vaccine vectored by the thermostable pox viruses are high even without considering the potential use of these viruses as vectors for a multiplicity of disease immunogens. Besides the elimination or at least the reduced need to maintain the cold chain in vaccine storage and transportation, it is claimed that the vaccines could be produced at greatly reduced costs. Before any benefits are realised, however, the whole concept of vectored vaccines will have to gain wider acceptance and the current widespread rejection of vaccinia as a suitable and safe vector will have to be overcome. There is already much activity in developing alternative animal poxvirus vectors including capripox and fowlpox which, unlike vaccinia, have no significant pathogenicity for humans.

Economic calculations for the thermostable vaccine have given internal rates of return of up to 0.22 even when used to Immunise relatively modest numbers of cattle.

Apportionment of Costs: Generally, the control of epizootics is perceived as a public good and paid for directly by the government although some countries have successfully introduced at least partial payment by cattle owners for these services. Recent experience in West Africa indicates willingness of producers to contribute to the cost of controlling rinderpest (de Haan and Bekure, 1990). The viability and sustainability of government veterinary services in developing countries would be substantially improved by taking advantage of this willingness on the part of livestock producers to pay for services.

Control of Foot and Mouth Disease As an Epizootic: Although capable of causing some mortality in traditionally kept livestock, many cattle owners fail to recognise foot and mouth disease (FMD) as an economically important disease. For this reason there is considerable resistance when they are called upon to pay for its control. In those countries which have an export market for cattle or their products, however, lost markets as a result of disease are readily appreciated and producer cooperation is easy.

In the case of FMD, regional agreement about its control would be desirable. Because of the multiplicity of FMD virus types and sub-types and the necessity to ensure monitoring of sub-type changes, the control of this disease will require regulation by government veterinary services. Diagnosis, quarantine, virus typing, vaccine production and coverage are all key steps in FMD control which require coordination and centralised authority. Many Latin American countries have lost this authority due to weakening of government veterinary services. Consequently the key steps are not coordinated for maximum effectiveness and often vaccine virus type is not related to virus type causing disease, vaccination coverage is haphazard and frequently ineffective. Until this situation is corrected, it will be biologically impossible to control, much less eradicate, FMD in these areas and any efforts will not be cost effective.

Not all countries place equal importance on FMD control. Emphasis is usually determined by the incentives of export markets for livestock and livestock products. Although it is technically feasible to control FMD, it is impractical in most countries where there are no incentives such as export markets.

Control of Newcastle Disease: Reliable figures on the importance of poultry in the diet or economy of various countries are difficult to obtain. This may have contributed to the slow progress in the control of poultry diseases. Newcastle disease constitutes the most important epizootic disease of poultry and is a serious constraint to the production of a cheap animal protein.

Where its control has been routinely practiced, use has been made of the standard vaccine, which like most other live viral vaccines, depends on the availability of a cold chain to maintain efficacy. Proof of the effectiveness of the thermostable oral vaccine and the successful promotion of its use would constitute an important step in increasing the effectiveness of Newcastle Disease control especially if it can remain stable in the wide variety of poultry feeds used in village situations.

The delay in the use of the oral thermostable Newcastle disease vaccine is puzzling and reasons for it ought to be investigated and impediments removed.

It is concluded therefore that practical methods for controlling epizootics do exist and beneficiaries are willing to contribute for their control or eradication. Policy backup is however required from governments to ensure that funds are made available to sustain disease control. This could be either through subvention of a higher proportion of the livestock contribution to gross national product or by a more rigorous programme to have beneficiaries pay for services either directly or through a system of cess.

Control of Enzootic Diseases of Economic Importance

In this category are included tick-borne diseases, trypanosomiasis, dermatophilosis and a number of viral diseases many of which are arthropod-borne. Although their routine control has traditionally been within the public sector purview, there is little doubt that most of the benefits accruing from their control go to individual livestock owners. In the past, methods for their control were too costly or not readily available to individuals. However, emerging technology is increasingly simplifying control and avenues are opening which should make it more cost-effective to have control provided by the government on a cost-recovery basis or done by the private sector and paid for by direct beneficiaries.

Tick-borne Diseases: Tick-borne diseases caused by Anaplasma, Babesia, Theileria and Cowdria species are of major importance throughout the tropics and subtropics. For many years their control relied heavily on vector control through the use of acaricides. The expense of these chemicals and reliance on costly modes of application using dipping tanks or spray races have meant that combating these diseases in most developing countries was a community centred effort often borne entirely by the government. This has been especially the case where livestock improvement programmes have been based on introduced exotic animals. Under those conditions enzootic stability, which protects indigenous livestock from unacceptably high mortality, could not be relied upon.

Factors which favour more cost-effective tick-borne disease control include:

• Development of pour-on acaricide formulations which do not need large volumes of water, a commodity which has in many cases been limiting. At the moment there are few such formulations and perhaps because of limited use they are still expensive. In East Africa their best use has been found to be under conditions of severe trypanosomiasis and tick-borne disease challenge (Omukuba, personal communication, 1989).

• Development of vaccines against the pathogens. In the past, these have been rather bulky and inconvenient to administer but have still represented a distinct advance over exclusive reliance on acaricides. Anaplasma and Babesia vaccines have been mostly attenuated strains of the organisms in whole blood with the many problems this entails, ie. bulk, short shelf-life, maintenance of the cold chain, potential danger from viral contaminants, etc.

While these are still in use in some countries, attempts have been made to develop blood-free vaccines. In babesiosis these have been derived from Babesia spp. culture (Ristic at al., 1981; Kuttler et al., 1982; Montenegro-James et al., 1987). More promising has been the research carried out to isolate the Babesia surface antigens that are responsible for protection (Wright et al., 1983; Goodger et al., 1987; Waltisbuhl et al., 1987).

Similar drawbacks are encountered in the use of *Theileria parva parva*, *T. parva lawrencei*, *T. parva bovis* and Cowdria vaccines currently available and administered through the infection-and-treatment method. The problem of extraneous contaminants is somewhat less with *Theileria annulata* with its tissue culture derived vaccine. The same can probably be said about *Babesia bovis* vaccines which have been developed from organism culture supernate.

- Promising advances on vaccines against the tick vector. Most advance has been made against *Boophilus microplus* the vector of anaplasmosis and babesiosis in a very large proportion of the developing world. When perfected this will constitute a major advance. It is considered a matter of time before similar advances are made against the other single and multi-host tick species.

- Government financial crisis which have curtailed funds for routine dipping services. Free or heavily subsidised government programmes for the control of tick-borne diseases have worked fairly well before the severe economic problems of recent years and the effect these have had on the availability of operational funds. Communal dips have been among the first casualties.

As the level of funding has declined, maintenance of the required acaricide strength and general supervision of correct dipping practices by public service employees has lapsed.

A measure of the utility of tick control in a given community has been the degree to which the operation of dips has been effectively taken over by the beneficiaries or the government has been able to continue running communal dips on a cost-recovery basis. When the main reason for using dips has been coercion through legislation and where the predominant breeds have been indigenous, farmers have often quickly discovered that not dipping their animals resulted in no substantial economic loss. Such systems are therefore essentially not cost-effective from the point of view of the farmer and should be re-examined or abandoned.

Tse-tse and Trypanosomiasis Control: Although it occurs in areas on other continents, trypanosomiasis is of special importance in Africa where its presence is associated with the tse-tse fly and where the disease in livestock co-exists with human sleeping sickness. The disease tends to cover large areas coinciding with the tse-tse habitat. For these reasons control of trypanosomiasis has traditionally been in the public domain. Routine control methods have been:

Vector control through:

- Insecticide sprays using fixed wing aircraft and helicopters by which large areas have been effectively cleared of the vector.
- Bush clearance was at one time extensively practiced and through which large areas were cleared of vegetation.

Environmental concerns and high costs have worked against this method of control.

- Slaughter of wild life was extensively practiced in the past in some countries with the aim of starving tse-tse out of given areas. This would be internationally unacceptable now.

- The use of tse-tse attractant baited traps and insecticide-impregnated screens as well as application, as pour-ons or in dips, of certain insecticides directly to animals at risk are emerging technologies which have lowered the cost of tse-tse control close to the levels where individual livestock owners can control the disease on their own.

- Maintenance of tse-tse control status usually requires some system of planned land use if reinvasion is not to occur.

Many people have questioned the desirability of opening large tracts of land to livestock through the control of tse-tse and the possible consequences of land degradation. Because of the need to encourage efficient utilisation of land, it is inevitable that people will eventually move into these areas as long as there are no effective measures of human population control.

Clinical Interventions

In many countries of the developing world clinical services have traditionally been provided by the government, often free of charge or heavily subsidised.

Factors which have contributed to less public sector financing of clinical veterinary services have included:

- Changing attitudes towards the view that clinical work is a service whose beneficiaries are individuals and therefore individual owners should be made to pay for it.

- Growing awareness by governments and supported by actual experience, that beneficiaries are willing to pay for effective services.

- Competing demands on government resources forcing higher priority to be given to services that have obvious public good implications such as the control of epizootics and veterinary public health.

- Great increase of veterinary staff establishments beyond the level at which governments can adequately and effectively utilise them.

As a result of these changes, a number of evolutionary developments with implications for greater cost-effectiveness are taking place. Among these changes are:

- Movement towards full cost recovery by the governments where these services are still provided through the public sector.

- The evolution of alternative ways to deliver these services, including:

  — Greater use of farmer groups/societies or cooperatives as employers of veterinarians who, besides clinical work, may manage artificial insemination services as well as herd health programmes.

  — Encouragement of private veterinarians to provide these types of services. Consideration is also being given in some developing countries to have such practitioners undertake contract work from the government for such duties as compulsory vaccinations.

Control of Helminthiasis

Tropics and subtropics offer ideal environments for helminths. Despite this and the well documented economic importance of endoparasites, their control is accorded low priority in developing countries. A factor in this neglect is no doubt their insidious nature in the more valuable adult animal.

Another, probably more crucial reason, is the lack of progress in the development of effective vaccines and the overwhelming reliance on the use of imported, and therefore expensive, anthelmintic drugs. Where anthelmintics are in extensive use, resistance often develops and laborious techniques are required to determine the degree of resistance and possible alternative drugs. Although many species of roundworms and flatworms are important in developing countries, perhaps the most devastating are *Haemonchus contortus* and Fasciola species.

In the tropics where the value of pasture rotation appears to be limited, reliance will continue to be placed on the use of anthelmintics, the administration frequency of which will have to be established in each given situation. The selection of the drug to use will often be an important consideration and facilities may need to be established on a regional basis for testing against resistance to given classes of anthelmintics with a view to using the cheapest, narrowest spectrum product needed to control the important helminth. This will guard against the tendency, widely promoted by drug companies, to utilise anthelmintics of ever increasing broad spectrum. There may be merit in giving attention to emerging initiatives in a number of countries to investigate the usefulness of indigenous vermifuges derived from local plants.

Many research groups in various parts of the world are now engaged in tests aimed at developing a vaccine against haemonchosis. If initial promise is borne out, success in this will have a profound effect on the cost-effectiveness of controlling the most important internal parasite affecting ruminants.

Operation of Veterinary Clinical Services

After years of debate and reflection, there is now relatively little doubt about the desirability of having owners pay for clinical services rendered to individual animals. This especially applies for afflictions which do not pose an immediate threat of spreading disastrously to other livestock.

Spurred on by the inadequacy of funds to finance urgent needs, such as the control of epizootics, governments are now increasingly willing to try a number of alternative ways of providing these services. Alternative mechanisms which are being supported by international donors and implemented in various African countries. These schemes employ various mechanisms for cost recovery of veterinary drug or vaccination distribution, delivery of veterinary services by pareveterinarians, private practioners or herder associations (de Haan and Bekure, 1990). Evidence to date is that in the face of full cost recovery, drug availability increased (CAR), drug purchases by poorer livestock owners increased (CAR) and rinderpest vaccination coverage was not adversely affected (Burkina Faso, CAR, Mali and Mauritania).

Services in Support of Disease Control

Routine disease control depends heavily on two main supports, namely, a diagnostic service and a source of vaccines. Laboratories for these are costly to establish for an individual country. Left to each nation to develop laboratories using its own resources, most countries would hesitate to do so. Such facilities are however almost invariably established and, usually for a limited period, operated with donor funds. Questions of cost-effectiveness are a rare consideration.

Table 3: Donor Projects Promoting Alternative Delivery of Veterinary Supplies and Services in Sub-saharan africa

Country	Donor(s)	Main Focus
Burkina	FAO	Village poultry/small ruminant vaccinations
CAR	IDA/IFAD/ EEC/FAO	Drug distribution and herder training
Cameroon	IBRD/IFAD	Drug importation/distribution & vet practice
Chad	GTZ/EEC/ IDA/ADB/FAC	Veterinary practice: vets and paravets
Cote d' Ivoire	CCCE/GTZ	Village pharmacies and paravets
Ethiopia	IDA	Paravets through service co-operatives
Guinea	IDA/CCCE/ FAC	Drug inputs and veterinary practice
Kenya	IDA/IFAD/ OPEC/EEC	Veterinary practice and cost recovery
Nigeria	IBRD/EEC	Studies
Mali	IDA	Paravets through herder associations
Senegal	IDA	Paravets through herder associations
Somalia	IDA/GTZ	Paravets and cost recovery
Sudan	IDA	Veterinary practice
Uganda	IDA	Veterinary practice
Zaire	IDA/FAC/ CIDA	Paravets through herder associations

Diagnostic Services

Epizootic Diseases: Many developing countries, large and small, have attempted to establish diagnostic laboratories for the major epizootic diseases, ie. rinderpest, FMD, Newcastle disease and pleuropneumonia of cattle and goats. However, only a few have been able to maintain them. Fortunately, with the encouragement and assistance of inter-national organisations, principally FAO, a few of the stronger national laboratories have assumed the role of regional diagnostic laboratories for the most important epizootic diseases.

A large proportion of the diagnostic work for FMD continues to be done in laboratories outside the developing countries because of what would appear to be non-technical reasons. In addition there have been moves over the past decade to standardise diagnostic reagents and techniques for epizootic as well as other important diseases. There are also continuing attempts to have, carefully positioned in various developing countries, truly international well equipped and staffed diagnostic laboratories.

These would be in addition to the existing regional collaborating laboratories which still remain the property of the host country.

Non-epidemic Viral and Bacterial Diseases: Confirmatory diagnosis of both viral and bacterial diseases remains a difficult problem where culturing and the production of reagents and use of more sophisticated techniques is required. At best such diagnostic work is done in centralised laboratories which results in considerable delays in obtaining results. Often laboratory reports are received long after the problem has passed. More effective diagnostic services will require greater investment in research aimed at developing simple but highly specific field tests. These may employ standardised sera/ monoclonal antibodies and corresponding antigens. Technology for their preparation exists and is becoming more accessible to developing countries. Their preparation however requires substantial investment in equipment and expendable supplies.

The establishment of regional diagnostic laboratories have been previously suggested by FAO. There is now a collaborative programme between FAO and OAU/IBAR to establish such a system. This will help in ensuring ready availability of standardised diagnostic reagents. A mechanism will however need to be developed to ensure that these are sustainable in the long run. Centralised mass production would help in reducing costs and therefore making them more affordable.

Sustainability must be an overriding consideration in running centralised laboratories for reagent production. For this reason ways have to be found to ensure a system of cost recovery. Alternatively, committed long term regional member country contributions to their operation will be required. Such commitments must include provision of a certain amount of foreign exchange allocation.

Diagnostic advances have an important bearing on tick-borne diseases control. Traditional diagnostic techniques using microscopy to examine blood, lymph node or brain smears are notoriously laborious and inaccurate. An important feature in current attempts to improve cost-effectiveness of tick-borne disease control is the development of highly precise diagnostic methods to ensure that the still expensive vaccines are employed only where indicated.

Vector-borne Protozoal/Rickettsial Diseases: Lack of sensitivity has been a serious drawback for the antigen detection techniques which have been traditionally used for Babesia, anaplasma, Theileria, Cowdria and trypanosome diagnosis. Both sensitivity and specificity have been greatly improved by new monoclonal antibody based antigen capture ELISA tests and DNA probes. For a number of these, diagnostic kits are already available for evaluation. These tests still depend on a considerable degree of sophistication and generally have to be done in central laboratories. They are useful for survey purposes but would probably not be cost-effective for routine use. Greater usefulness will require further simplification to make them directly applicable at the field level.

Vaccine Production

Because of understandable desire to ensure protection against epizootics, many governments in developing countries have established vaccine production laboratories. There are a number of alternatives which can be pursued to improve production efficiency which include the following:

- Vaccine production through parastatal organisations: To cater for national needs, there is a growing tendency in developing countries to put vaccine production under specially created parastatal entities. The combination of fickle government funding, lack of market orientation and limited national markets, do not hold much promise that this mode of organisation will enhance cost-effectiveness or substantially.

- Vaccine production under the umbrella of a regional organisation such as OAU: This would have the advantage of political support and the potential of a large regional market through which economies of scale could be realised. Such an organisation would, however, need to overcome management problems which have vitiated similar ventures at regional and subregional level in the past. An insurance to success would be to make production laboratories self-sustaining.

For this a business management approach such as is being promoted in Botswana would be required. Careful note would therefore have to be taken of organisations of similar nature which have succeeded and the necessary lesson learned.

- Vaccine production through commercial vaccine enterprises: Despite their distance from third world markets, vaccine companies in Europe and North America have continued to produce vaccines whose landed prices within the developing countries are cheaper than locally produced vaccines. This has been attributed to relative production efficiency, superior technology, management and marketing excellence and a global market. A factor constraining the establishment of more vaccine plants by international pharmaceutical companies in developing countries is the difficulty encountered in moving vaccines across national frontiers from a central laboratory. Regional organisations such as OAU have had remarkable success in mobilising vaccine transfers throughout Africa. Development of a suitable linkage mechanism between such regional organisations and commercial vaccine companies, which need not be entirely foreign to the region, could be one way of capitalising on the relative advantages of both entities.

Delivery of Vaccines at Farm Level

As has been discussed, control of epizootic diseases through regional mass vaccination campaigns has been effective and has saved national herds from such diseases as rinderpest and CBPP. As long as these epizootics exist international efforts through governments would appear to be the only effective means of vaccine delivery. As these diseases are controlled, the maintenance of the immune status might devolve to national veterinary departments which would decide whether this can be done more cost effectively directly by them or through the emerging private practitioners via contract arrangements.

These decisions will only be possible if the veterinary profession as a whole and national veterinary departments in particular take greater interest in the application of economic principles. It is noteworthy that several African countries are considering the establishment of veterinary epidemiology and economics courses in their national universities and at least one has already instituted such a course. This is a trend which needs to be encouraged and supported.

Importation and Distribution of Drugs

In contrast with vaccines for which the developing countries now have a great deal of indigenous technology, the third world is almost completely dependent on the import of therapeutic drugs. In Sub-Saharan Africa, drug importations are carried out, with few exceptions, directly by government or government parastatals. This mechanism has been notorious for providing poor distribution and availability of veterinary drugs at the farm level. Farmer demand for effective drugs has frequently been met by smuggled drugs with all the attendant problems of lack of consistent supply, questionable handling practices, high cost and loss of government tax revenue. Despite these limitation, smuggled veterinary drugs in many countries have been the primary supply and distribution method.

"Revolving funds" for drug purchase and delivery are one method used to liberalise drug supply in numerous African countries. Overall return of funds ranges from 30 to 50 percent when operations are decentralised and collections are made through vaccination posts. The most successful scheme (for example, in the CAR), which returns 85 to 90 percent of expected funds, utilises herder associations to collect the money and a double accounting system of accountability is used. Veterinary drug distribution through private pharmacies or farm supply/fertilizer outlets is used in Asian and Latin American countries. Drug availability seems to be better than in Africa. However prices may be exceedingly high due to the inherent high cost of small dose packaging and low turnover. Where veterinarians are actually seeing and diagnosing clinical disease and prescribing drugs for farmer's animals, turnover is faster. Additionally if veterinarians think in terms of "community epidemiology", all at-risk animals receive treatment or preventive and consequently lower cost, multi-dose packaged drugs are stocked and prescribed. These two conditions, active delivery of clinical services and treatment of all at-risk animals, tend to reduce the per treatment cost, are generally more efficacious and are medically more sound.

Management of Animal Health Services

Traditional Approaches

In many if not most developing countries, veterinary services were established to provide diagnostic and control services for unfamiliar tropical diseases which puzzled colonial settlers or to fight diseases introduced by them to new environments. The number of veterinarians was for many decades low and a tradition was established to employ all of them within the government. This continued to be the case even after the major epizootic diseases were brought under reasonable control and the functions of veterinary departments became more diversified. The policy of guaranteed employment for all qualified graduates in these increasingly complex and diversified veterinary departments has been brought into question. The debate continues about which functions should be priority undertakings of the public sector and which should be left for management by the growing private sector of veterinarians.

Even with increasing numbers of veterinary graduates, and in some respects because of it, the delivery of veterinary services to the majority of livestock raisers is declining in many developing countries. The increasing employment of veterinarians in the government services over the last thirty years has outpaced the rate of increase of funds to run veterinary departments. As a consequence, the percent of funds devoted to non-salary operational costs (in absolute terms) has dropped by approximately one third in West African countries and by two-thirds in Kenya. This situation points out the fallacy of relying on livestock unit:personnel ratios to plan the delivery of veterinary services. Even if ratios of 1000:1, or probably even 100:1, were achieved, veterinary services would not be effectively delivered if supplies, transportation and other in puts were not provided to facilitate the actual delivery of goods and services to the livestock.

Table 4: Main Contributor to Non-availability of Services

Region	Percent Budgetary Allocation		
	Year	Salary	Non-salary
W. AFRICA	' 60	64	36
	' 76	75	25
KENYA	' 74	40	60
	' 81	70	30
	' 89	80	20

Government Management of Animal Health

Management of Notifiable Disease Control: There is little doubt that animal diseases with epizootic or zoonotic potential are of state interest and their control should be a public responsibility. This responsibility includes legal obligation for farmers/animal owners to report suspected disease occurrence to the government and for the veterinary authorities to give prompt confirmatory diagnosis as well as imposing such control measures as are appropriate for containing disease spread. Apart from vaccinations, controls could include quarantines or animal disposal.

There are legal penalties for non-compliance. While ultimate responsibility for enforcing the control of such diseases, clearly rests with the government veterinary authorities, there are measures that the state can take under certain conditions, which may improve cost-effectiveness. Such measures may include:

- Where there are private practitioners, either self-employed or engaged by farmers' groups or cooperatives, preliminary reporting, vaccination and primary supervision of stand stills could be contracted to them provided the government still retains overall responsibility.

- In the case of compulsory vaccinations, legal contracts could be let. Efficiency and reliability are facilitated where independent means for assuring vaccination coverage, such as through serological surveys, are available.

When combined with other measures, such as the withdrawal of unfair competition in clinical practice by government veterinarians, contracts of the type described above could be an incentive for veterinarians to become self-employed.

Veterinary Public Health Management: Management of abattoir functions including the assurance of humane slaughter of animals and meat inspection is governed by individual national legislation and generally is in accord with international zoosanitary codes.

This and other food inspection functions, depending on circumstances obtaining in a given country, may be under veterinary control. Efficient management of these functions presents considerable problems because outside the large urban areas, which have centralised slaughtering and food processing facilities, rural facilities are scattered and difficult to police.

Here again, provision of these types of services could be based on cost recovery mechanisms in the form of user fees to butchers, meat markets and food processors. These fees would likely be passed on to the consuming public as a charge for their public health.

Education and Training Needs of Animal Agriculture

The availability of correctly trained and skilled manpower is one of the most critical requirements in developing a livestock sector. However, it is only one of the constraints that limit animal production in developing countries. The number of skills that are needed are many and varied. Certainly an adequate supply of skilled manpower will not of itself ensure development, especially if it is wrongly deployed or cannot be deployed at all.

In many countries there may even be a surplus of skills in certain sectors but that these skills cannot be used because there is no money to pay for the effective use of the trained personnel. A classical example of this exists in Egypt. In this country there are 5 veterinary schools producing a large number of veterinarians each year. The result is that this country has a total of 16,500 veterinarians of whom, all but 800 are in government service. These veterinarians service a comparatively small animal population of 2 million cattle and 2.5 million buffalo. This compares with the situation in the United Kingdom where there are only 8,500 veterinarians, 7,000 of whom are in the private sector. Only a small number of these would service the needs of the 12.5 million cattle found in this country. Many of the remainder would earn their living looking after the welfare of "companion animals" such as cats and dogs.

Clearly all the veterinarians found in Egypt cannot be gainfully employed and certainly the high number does not mean that animals in Egypt are better cared for than those in the UK. Also the provision of salaries of the veterinarians in the public sector will place an unnecessary burden onto the expenditure of the Egyptian Government. This one example shows us that more does not mean better and that the provision of trained personnel will not necessary solve a country's development needs. In fact, too much training may in fact be counterproductive. Only if other constraints are removed can spending on training be a useful investment.

In the above example only veterinarians are considered, but many other disciplines are needed to ensure that sustainable animal agriculture can be practised. What are the constraints to animal production in the tropics and what type of training is needed to overcome these constraints. These factors will be dealt with in the next section, whilst in the final section the training of veterinarians and para-veterinarians will be looked at in detail and the ways in which their numbers and training can be modified to meet the need of individual countries.

Factors that Limit Animal Production in the Tropics

The factors that limit animal production in the tropics include the following:

- the direct and indirect effects of climate;
- the genetic merit of available livestock;
- the quality and seasonal availability of food supplies;
- the health on animals both on the farm and in the region;
- the level and type of management with particular reference to the affect of religious and social attitudes; and
- the availability of credit, particularly to poor farms.

Skills that are Needed to Overcome these Constraints

The relative importance of each of the above constraints will vary from country to country but the overcoming of these constraints will need the input of a number of specialists, examples of which include animal physiologists, breeders, nutritionists, forage/range agronomist, veterinarians, sociologists and economists.

This list is not, of course, exhaustive but does give a fair idea of the many different disciplines in which training is needed. It will not be possible in this short presentation to cover all disciplines mentioned. What we will do is to use the example of veterinary medicine to investigate the problem and to see what types of training are needed, where they are available and how effective training is in ensuring development.

Training and Sustainable Development

The concept of sustainability is one of keeping a system going. This aim might not be desirable, if this system is being sustained using a large number of external inputs. For example, it may be possible in many countries to sustain intensive poultry production using imported feeds, birds and vaccines. Such an enterprise would need trained personnel to run it such as veterinarians, nutritionists, possibly geneticists in addition to skilled manual workers. However, would such trained labour be justified when the effect of such a programme would be to reduce the total of food available in a country?

This type of system can rarely be justified in developing countries (Smith, 1990).

Therefore when identifying training needs it is important to identify the system into which the trained personnel are to be slotted. In most cases the system should be one that will sustain itself with the minimum of outside intervention and the training received by the personnel should reflect this concept.

Nomadism in certain areas of the world, is or was, a sustainable form of animal agriculture until outside forces caused it to become unsustainable. The main factor causing unsustainability is normally population pressure but ill conceived education or training programmes have sometimes made the problem worse. For example the conventional education of children of nomads would certainly remove them from the labour pool whilst they are of school age and might alienate them completely from this system of production. Similarly training technicians to drill bore holes may lead to chronic overgrazing in the area around the boreholes. Therefore it must not be thought that training is always beneficial it can as we have seen make matters worse.

Inability of Government to use these Skills because of a Lack of Operation Funding

Another problem that faces planners in the third world is, having trained skilled manpower, how can this manpower be used in developing countries. Most of the available manpower is deployed in the public sector and in many cases the public sector cannot afford to pay for skilled labour.

Table 5: Deployment of Veterinarians in Certain Selected Countries

Country	Employment by government	In gov't laboratories	Private sector	Total
Egypt	11,700	3,500	800	16,350
Nigeria	900	274	274	1,621
UK	483	771	7,011	8,489

In a survey carried out by Denis Fielding and one of the authors (AJS) in 21 countries in East and Southern Africa in 1986 we found that in few, if any, of the research stations and Universities devoted to animal production made any real impact on the output from this sector. One of the main reasons for this lack of impact is that practically all the money (between 85 and 95%) of the expenditure goes to pay the salaries of the personnel, whereas in a well balanced organisation only about 65% should be used for this purpose. The situation has become worse over the years due to the worsening financial situation of third world countries and even relatively well off countries, such as Zimbabwe, are now spending over 70% of their relevant budget on salaries. At that time (1986), one of the major fears of the then Director of Zimbabwe Department of Research and Specialist Services was that the Government would make him take an increased quota of students graduating from the local University. If this was to happen he feared that there would be no money for actual research because all the available funds would have been used up on salaries. Such a situation has been reached in Tanzania where the Government practice is to employ all graduates in the public sector.

Where such a policy is pursed salaries are low, often well below a living wage, and not only do the staff not have money for work but they have to take on additional jobs in order to survive. Under such conditions it must be asked what value these qualified people are to their country and would their country's needs be better served by fewer graduates.

Aspect of Training Needs of Animal Agriculture

We have chosen to focus on veterinary training because it is the type of training which is most clearly defined and it is the one on which there is the most factual information.

Veterinary training requirements to fall into three categories as follows:

- Graduate Training.
- Post Graduate Training.
- Technician Level Training.

All these areas shall be briefly considered, but in the view of the authors, the last category has the greatest need.

Graduate Training

The need for veterinarians in different countries and regions depends on the type of livestock industry. Botswana has one of the most effective government veterinary services in Africa and can be used as an example of optional staffing levels.

Botswana has 30 veterinarians, mostly employed in government veterinary services (FAO, 1988) and a total of 2.4 million livestock units (LU) of which cattle represent 2.3 million LUs. Several factors combine to simplify the provision of effective veterinary services which are:

- flat terrain,
- a livestock population dominated by cattle which are, predominantly extensively grazed, and
- relatively few disease problems.

Thus veterinary services are effectively provided by a small professional cadre supervising field staff who carry out most of procedures (vaccinations, castrations, dosing, quarantine, inspections, etc.).

Using Botswana as a model, it can be argued that its ratio of 80,866 LU's per veterinarian is the optimum for effective veterinary services. In other situations (e.g. a more intensive livestock industry, more difficult terrain, greater diversity of animal husbandry practices) the optimum ratio for effective veterinary services would be lower.

Schöner and Meyer (1971) analysed the number of veterinarians in relation to domestic animal population throughout the world. Using the Botswana model, an arbitrary ratio of 100,000 LU per veterinarian has been used as the minimum requirement to deliver effective essential veterinary services and countries with ratios greater than this (i.e. more than > 100,00 LU per veterinarian) are shown.

From this crude, arbitrary analysis it can be seen that in 1971 the greatest shortage of veterinarians was in Africa, particularly, francophone West Africa, with most of South America and Asia being reasonably provided with veterinarians.

Currently there are two veterinary faculties in francophone sub-saharan Africa, namely at the "Universite Nationale de Zaire" in Lumbumbashi and the "Ecole Inter Etats des Sciences et Medicine Veterinaire de Dakar" in Senegal which encompasses 13 African countries (Brandt and Geerts, 1987).

By contrast, countries in anglophone sub-saharan Africa have strived to develop their own national veterinary schools. Hence, since 1971, the numbers have trebled from five to 15. Thus the question arises which is the best approach, francophone or anglophone? The authors have only a limited experience of the francophone regional approach to veterinary training but have had some experience of the

anglophone system. Veterinary schools are very expensive to establish
and run; for example, it is well documented in the UK that veterinarians
are the most expensive graduates to produce. Hence once a country
has invested in a national veterinary school, it runs the risk of producing
graduates in excess of its needs to justify the investment. The problems
associated with excess training and over production of graduates has
been mentioned above.

Table 6: Countries with more than 100,000 Livestock Units (LU) per Veterinarian

	Country	*Ratio of L.U.: Veterinarians*
Africa	Guinea	<1,000,000
	Mauritania	500,001–1,000,000
	Mali	"
	Upper Volta	"
	Ethiopia	"
	Nigeria	200,001–500,000
	Chad	"
	Nigeria	100,001–200,000
	Senegal	"
	Tanzania	"
	Cameroon	"
	Lesotho	"
	Morocco	"
South & Central America	Haiti	>1,550,000
	Paraguay	150,001–300,000
	Honduras	100,001–150,000
	Nicaragua	"
Asia & Oceania	Afghanistan	300,001–700,000
	South Yemen	200,001–300,000
	Nepal	100,001–200,000

Source: Schöner and Meyer (1971).

Table 7: Veterinary Schools in Anglophone Sub-Saharan Africa

Country	*1971**	*1990*
Kenya	1	1
Nigeria	2	5
South Africa	1	2

Sudan	1	1
Somalia	-	1
Ethiopia	-	1
Uganda	-	1
Tanzania	-	1
Zambia	-	1
Zimbabwe	-	1
Total	5	15

* From WHO (1973).

If new veterinary schools are under consideration to supply veterinarians for these countries, it is strongly recommended that the regional approach to training, as prevails in francophone sub-saharan Africa, should be investigated.

Irrespective of the location of a veterinary school, the curriculum has to cover a similar range of disciplines (physiology, biochemistry, anatomy, pharmacology and medicine, etc.), although emphasis in the clinical areas varies from country to country and region to region depending on needs.

It is impossible to give an optimum staffing level for a veterinary school, but in order for a school to be effective, it must have a minimum critical mass. The academic staffing levels in British and Irish Veterinary Schools range from 55 to 78 (average 63) and these are complemented by research staff and honourary fellows (Royal College of Veterinary Surgeons, 1990).

It would be difficult to see how a veterinary school could be effective with less than a critical mass of about 60 professional staff carrying out both teaching and research. In order for veterinary schools to be cost effective, the ratio of students to staff need about 300 students which for a 5 year course means about 60 graduates per year. This is probably over ambitious for most developing countries and a more realistic figure is probably about 20–30.

Assuming 20–30 graduates a year is a reasonable target, how many veterinarians are required to justify this level of production? In the UK, exhaustive analyses of this type of question in relation to British Veterinary Schools have been carried out in recent years. In 1988, 8,489 veterinarians were employed in the UK (FAO, 1988). The production of approximately 300 graduates per year is adequate to sustain this. Consequently by simple arithmetic; 20–30 graduates a year would sustain a cadre of 565 to 850 veterinarians. Thus if the number of veterinarians required for a country is significant less than this, it must be questionable if a national veterinary school is justified. The number of veterinarians needed to maintain a government Veterinary Service is estimated by us to be one veterinarian per 100,000 Livestock Unit.

Post Graduate Training

In the authors' view the requirements for post graduate training are adequately supplied. This view is not necessarily shared by others, including potential candidates for post graduate training.

To anyone who has travelled and worked in developing countries, the perceived clamour for further training and post graduate degrees is frequently at odds with the low level of application of basic graduate training and education: the "I need to do a PhD" syndrome is well known to us all. Post graduate veterinary training has been reviewed by several authors and some programmes available are summarised.

To the aspiring MSc or PhD post graduate veterinary student, the world is his oyster if he can secure the necessary funds. The greatest problems arise in ensuring that such a student embarks on relevant and appropriate training.

There is always the risk that the candidate with sponsorship will register in "a course" to utilise the funding, even if the course is not strictly what is required. Ideally, those responsible for organising and administering post graduate education should respond to needs dictated by livestock and veterinary departments, donor agencies, etc.

In reality, the needs are rarely defined and the educationalist have to assess these needs themselves. At the CTVM, we do this by maintaining a wide range of links with overseas organisations an individuals and couple our "intelligence" with the more stereo-typed-up-dating of information from the scientific literature.

There are clear advantages of training MSc students out of their own country. This type of training enables them to mix with students from other countries that often have similar problems to their own. It also enables them to see their own country's problems from outside and therefore to understand them better.

Table 8: Estimated Numbers of Veterinarians, Cattle Stock and Cattle/ Veterinarians Ratios in Countries in Central and South America *

Country	Number of Veterinarians	Cattle Stock (millions)	Cattle/Veterinarian Ratio
Central America			
Belize	10	0.05	5,000
Costa Rica	430	2.55	5,937
Cuba	3,611	6.40	1,772
Rep. Dominicana	400**	2.42	6,050
El Salvador	218	0.92	4,261
Guatemala	425	2.58	6,087
Haiti	5	1.35	270,000
Honduras	110	2.50	22,800
Jamaica	39	0.32	8,230
Mexico	18,000**	37.45	2,080
Nicaragua	69	1.89	27,391
Panama	406	1.42	3,504
South America			
Argentina	9,202***	54.80	5,955
Bolivia	505	5.85	11,584
Brazil	24,420**	134.50	5,507
Chile	2,210	3.40	1,538
Colombia	5,000	21.93	4,387
Ecuador	2,000	3.37	1,689
Guyana	23	0.14	6,086
Paraguay	828	6.40	7,637
Peru	2,408	3.90	1,619
Uruguay	1,500	9.94	6,632
Venezuela	6,700**	12.48	1,862

* FAO (1985, 1985b).

** Estimated figures from the faculties consulted.

*** de Diego (1985). After Campero (1987).

Information on available training programmes can be difficult to locate. Thus one of the authors (AGH) on a recent training mission to South America experienced great difficulty in identifying suitable

training programmes for Spanish or Portuguese speakers employed
in veterinary laboratories. An up to date register or directory of post
graduate training programmes would ease many of the problems
encountered in organising relevant training and it is strongly
recommended that if such a document is not available, it should be
published by an appropriate agency, e.g. FAO.

Table 9: Post Graduate Programmes for Veterinarians in Developed and
Developing Countries.

Region	Country	Type of Programme and Number of Schools/Institutes Available		
		Short Course	MSc	Phd
N. America *	Canada	1	2	2
	USA		14	14
C. & S. America (*)	Argentina	1	1	
	Brazil		20	6
	Chile		2	2
	Colombia		2	-
	Venezuela		2	-
	Mexico		7	1
Africa *	Ethiopia	1		
	Kenya	1	1	1
	Nigeria		3	3
	Sudan			1
Asia *	India	2	8	7
	Thailand		2	2
	Malaysia		1	1
	Indonesia		2	2
Australasia *	Australia	2	5	5
	N. Zealand	1	1	1
Europe +	U.K.	2	5	5
	France	1	1	
	belgium		1	
	Germany	2	1	
	Holland	2	1	

* From Campbell (1984)
(*) From Campbell (1984) and Campero (1987)
+ From Edelsten (1984).

In many instances, and for many individuals, the most appropriate training at post graduate level is the short course designed as a refresher or up-dating course. These can range from a few days to a few weeks and involve one individual on attachment, or a more general structured arrangement with delegates and invited tutors. These tend to be organised in the developed world and thus tend to be limited to the fortunate few who can find the necessary sponsorship. Greater consideration should be given to running short courses "in situ" as these would reach a wider audience and be available to many more delegates.

Technical Level Training

This very important area is the most difficult to define. Most countries have some arrangement for training field and laboratory technicians but the quality of the training can range from the abysmal to the excellent.

It is usually possible to assess the quality of the various graduate and post graduate programmes and act accordingly, but this is not the case with technician training. There is also the question of how many technicians are required. Once again, using Botswana as a model, some indicators can be deduced.

Botswana has 1,418 veterinary auxiliary staff (FAO, 1988). The majority work in the field, but others work in laboratory services, quarantine camps, abattoirs and livestock advisory centres. They are supervised by 30 veterinarians, i.e. approximately 47 auxiliary staff to each veterinarian.

Clearly this ratio might be inappropriate in veterinary services where greater direct professional intervention is required, but it could be argued that an arbitrary ratio of 40 to 50 auxiliary staff to one veterinarian is a good baseline for planning staff numbers. Thus in Botswana, the ratio of veterinarians: technicians: L.U.'s is 1:47:1710. In other circumstances, farmers may be expected to attend to certain veterinary procedures themselves (e.g. dosing against helminths and castrations, etc.) and thus the ratio of technicians per L.U.'s may be greater than 1:1710 - say 1:5000. Likewise, in certain circumstances, veterinary procedures and the arbitrary ratio of veterinarians: L.U.'s of 1:100,000 may have to be reduced to a lower figure, say 1:25,000, in which case a more realistic ratio of veterinarians: technicians would be approximately 1:10.

The important point is to ensure that these ratios are estimated as correctly as possible, so that realistic staffing levels and training programmes can be drawn-up.

It is probably unrealistic to analyse the various training arrangements for technicians in different countries. What would be more practical is to give consideration to running training programmes in recognised reputable centres where the quality of the training could be controlled and standardised.

In order to alleviate limitations in languages of technicians, these programmes should be conducted in the major languages of the world (English, French, Spanish and Arabic) and be conducted in suitably located centres.

The need for such training is particularly great for laboratory technicians. In too many countries, because of the low level of training and skills of laboratory technicians staff, professional staff have to resort to carrying out routine technical chores, so preventing them from utilising their professional training and education to full advantage.

Strategies for Effective Extension Services to Guide the Advancement of Animal Agriculture in Developing Countries

Over the last decade most of the increase in animal products have been achieved through an expansion in stock numbers and not as a result of increases in animal productivity. With a declining resource base such a situation is clearly not sustainable.

The challenges facing agricultural extension in general, and the livestock sector in particular, are formidable. The increasing world population will continue to be dependent on declining resources to feed itself. More efficient production systems must be introduced, but not at the expense of the environment or resource base, and conservative, risk-adverse, traditional farmers must be persuaded to become more innovative. At the same time, this must be achieved within a framework of fluctuating commodity prices, erratic policy making and increasing financial constraints. A functional, efficient and cost-effective extension service is a major prerequisite if we are to develop long-term sustainable livestock production systems.

Throughout the world numerous different extension approaches have been tried in the attempt to improve farming practices. These approaches have included: the classic general extension service; commodity based services; the training and visit system; participatory services linked with producer associations; project driven services; the farming systems approach; and the educational/institutional (University based) extension approach.

As would be expected, livestock extension services throughout the developing world vary considerably in respect of institutional arrangements and resource allocation. For example, Malaysia continues to maintain a specialised livestock extension service which is both adequately funded and staffed with well trained officers. Agritex, the Zimbabwe extension service, is a good example of a country-wide service of relatively well trained and equipped general extensionists, supported by teams of subject matter specialists and veterinarians.

In Chile, extension work is undertaken by private consultants, selected on a competitive basis with funds provided through the National Institute for Agricultural and Livestock Development (INDAP). Unfortunately however, in many countries livestock extension is a service in name only. The reality remains that all too often livestock extension is restricted to providing limited services, such as, AI (rarely effectively) or animal health control programmes with the emphasis on specific diseases; rather promoting services or interventions aimed at improving overall animal productivity.

The trend in the provision of livestock extension has been away from the specialised, usually veterinary orientated, commodity approach, towards general, multi-disciplinary services. A notable exception is Botswana which has invested heavily in its livestock industry (but less so in livestock extension) in response to its favoured access to the european markets and, consequently, its ability to sell beef well above world market prices. The rationale, and it is correct, for dealing with livestock as an integral part of the general extension service is that the farm enterprise consists of series of interrelated activities. These activities are co-ordinated by the farmer utilising the resources that are available to him. Livestock are usually, except for specialised enterprises, an integral part of the farm and should therefore be treated as such. What is required therefore is an extension agent that is capable of taking a holistic view of the farm.

Within extension services livestock usually take second place to crop production. In a way this is understandable and is unlikely to change. Food and cash crop production are often economically more important to the national economy than livestock products, hence they receive a higher priority in the national development plans. Technical interventions applied to crops can have an immediate and quantifiable impact, in contrast, many of the interventions associated with improved animal production take a longer to fruition and are not simple to evaluate.

As far as livestock extension is concerned, the use of general extension agents does however raise a number of practical issues regarding how a multi-disciplinary extension worker can be provided with the sufficient range of knowledge and skills needed to be effective. The skills required for animal husbandry are very different from those associated with cropping and agronomy. For an extension agent to be effective they must have the confidence of the producer and that requires credibility - this is particularly important when dealing with livestock. Credibility comes with being able to show procession of knowledge and, more importantly, practical skills.

Constraints to Providing Livestock Extension Services

Many of the constraints affecting livestock extension apply equally to the overall extension service. To highlight some of these issues it is worthwhile examining what is expected of the extension agent. Ideally she/he will be a paragon, competent in a wide range of technical and social skills, having a knowledge of:

- of the technical subject matter in areas such as agriculture, animal husbandry, fisheries and home economics,
- the extension service, its organisation, operation, grants/subsidies and government regulations,
- the development of human resources through the participatory processes and group dynamics,
- communication skills capable of imparting knowledge and obtaining feedback, and
- teaching and extension methodologies.

Clearly this is an ideal, whereas the reality is that all too often extension agents are: under-trained, under-equipped, under-employed, under-paid and under-motivated.

Furthermore, the general extension worker has a multiplicity of duties to perform, including:

- the provision of technical advice covering a range of commodities, including animal production,
- assisting in the supply of inputs (seeds, fertilizer, breeding stock) and administrating/supervising credit, plus
- administrative, regulatory and statistical duties.

In attempting to undertake these duties they are often hampered by:

- a lack of appropriate knowledge and skills,
- a lack of appropriate extension messages and technical packages,
- a multiplicity of duties,
- maladministration, poor line-management and programming, often compounded by regular restructuring of ministries, provincial and local administrations,
- a lack of recurrent funding and logistical support (transport, travelling allowances, housing, basic equipment etc.), and
- a lack of technical support and the unavailability of inputs (credit, breeding stock, drugs etc.).

These are the same constraints that affect the service as a whole and can be broadly classified as internal and external constraints.

External Constraints

External constraints relate to the larger social, economic, administrative and political environment in which the extension service must work. Many of these macro issues have been discussed in the second plenary session on "Policy and Environmental Issues", however, we raise them again briefly as they concern the implementation of extension services.

Government policy usually dictates priorities and resource allocation within the agricultural sector and, as mentioned before, livestock production usually takes second place to the food and cash crop sectors. Government policies, however, do not necessarily reflect the priorities of the farmers and producers. It is not easy for an extension service to convince a farmer to adopt practices which are unlikely contribute to either increased income or fulfil a specific perceived need. Extension cannot coerce people to do things they do not wish to do.

Pricing policies have provided classic examples of government policy working against the local producer. Controlled farm gate prices, usually way below world market prices, are used to ensure a cheap food supply for the urban (politically active) populations. For example, the importation of subsidised milk powder has, in the past, ensued that many a domestic milk industry has failed to develop, as has a happened in the Philippines.

Combinations of educational and employment policies have lead some countries to build up vast public sector organisations in terms of manpower, yet completely stricken for recurrent funding with which to undertake extension duties. For example, Egypt now has over 15,000 veterinarians and a government committed to finding employment for all new graduates. It is ironic that underemployment is probably as large a constraint as the lack of skilled staff to agricultural development.

Internal Constraints

Lack of Appropriate Technical Messages. The shortage of technology is often cited as a major constraint to livestock development. This is not true. There is ample technology available ranging from the basic principles of animal husbandry through to improved nutrition based on recent understanding of the rumen function and sophisticated biotechnology, capable of increasing animal production. What has been lacking has been the capability to interpret when and where such interventions are appropriate.

Interventions are only appropriate if they are acceptable to the client and capable of satisfying their perceived needs. Two provisos must, however, be added:

- that interventions are technically, economically and ecologically sound, and
- that all needs as perceived by the farmer are not necessarily sound eg. a producer might feel that he needed to upgrade his stock with exotic breed - a course of action that could easily be totally inappropriate.

Fundamental to this question is an understanding of the farmers actual objectives for keeping animals. In commercial and semi-commercial operations the objectives are clear and constraints relatively easy to identify and correct. This is not necessarily the case regarding the small-holder sector. The primary outputs from animal production, such as, milk, meat, hides/skins and fibres may only account for 50% of the total returns to the farmer, and secondary benefits, such as, manure, draught power, capitalisation may be equally as important.

However, invariably, interventions originate from the top and are passed down to the producer and are often aimed at one particular sector, say milk production. The point is made that a 50% increase in milk production may only corresponds to only a 10% increase in overall returns to the producer, if milk represents 20% of the total animal products. These are clear cases were the objectives of the producers are not the same objectives as those of the authorities. Communication between the producer, researcher and policy maker is one of the fundamental principles of most extension systems. It is also the main area where systems have failed to live up to original expectations. The failure to involve the producer in the development of extension programmes has been a major constraint.

All too often we have seen technologies advocated, often with international assistance, that:

- are replications of sophisticated techniques imported from the developed countries which are inappropriate, expensive to implement and unlikely to achieve results or, worse still, have an adverse effect,
- ignore or underestimate local resources,
- fail to recognised the need for an integrated approach, for example, improved management and nutrition is a prerequisite for a breed improvement programme, as are policies to stimulate off-take for an animal health programme with the principle aim at reducing mortalities, and
- are appropriate but have failed to be accepted due to external factors, such as, pricing, inadequate marketing and processing facilities. Technologies cannot be viewed in isolation.

Lack of Trained Staff. The basic qualifications of the general extension worker varies from country to country and ranges from degree holders, diplomates from agricultural colleges to agricultural high school graduates and secondary school leaves. Some of these higher level training institutes may offer animal husbandry as an option, but generally coverage given to animal production is rudimentary. Partly, this is due to the fact that these institutions lack the necessary expertise and facilities required to provide adequate practical experience. The consequence is that the majority of agriculturalists have had little exposure to animal husbandry. There is clearly a need for strengthening both research and educational institutions and, equally important, their links with the extension service.

In-service training should provide the opportunity to narrow the gap between current level of competence of staff and that necessary to fulfil the duties required of them. Most countries have inservice training programmes implemented through, or supported by, a network of agricultural (some specialised) training centres. However, as far as training in livestock production is concerned it is usually provided on an *ad hoc* basis.

Without being too specific or deliberately negative, there are a number of inadequacies that can be identified regarding the provision of in-service training in animal production, these include:

- course contents that are technically inappropriate and bear no relevance to the extension messages/technical packages being promoted,
- training staff that lack expertise in animal husbandry and, as a consequence, give emphasis to the theoretical aspects and usually based on information taken from text books which, in turn, are not necessarily appropriate,
- usually there is neither the expertise or facilities available to undertake practical training,
- often the content is too sophisticated, when what is required are simple concepts and practical skills,
- there is little follow-up of participants once the have completed training, and
- the real danger that managers of training institutions are forced into playing a "numbers game" to fulfil pre-set targets in which becomes the objective, rather than providing appropriate training.

Lack of Programme Planning: National policies and objectives regarding livestock development are not always either well prepared or stated. All too often objectives derive from centralised autocracies and have little regard for variance in either agro-climatic conditions, farming systems - let alone the requirements of the actual producer. Defined policies and realistic objectives based on sound rationale are required at the national, provincial and local levels, if there is to be any chance of sustainable livestock production. There is clearly a role for small, autonomous extension services better suited and accountable to the client (producers).

Lack of Communication: Extension should be at the hub of a network of information, input and support services aimed at the producer. However, these linkages, especially in the poorer developing countries, are weak. This is particularly the case in the most important link between the farmer, extension and research. Communication should be bi-directional with the extension services acting as the principal intermediary between the producer and research, government and other interested institutions. In the case of animal production the role of the Subject Matter Specialist is critical in the communication process.

Inter-institutional communication is often sadly lacking. Research and extension services are often separated in different ministries. Where there do occur together, strong vertical hierarchies exist within divisions whilst horizonal links between them are weak. The Indonesian Ministry of Agriculture is a case in point where livestock development, extension, training, research, marketing, programme planning and many other departments all fall within the Ministry's mandate - the lack of co-ordination between them is a major constraint, certainly, for livestock production. Again, Indonesian is by no means the only example.

Lack of Resources: There is no doubt that a lack of resources are a constraint to all extension services. Within an organisation salaries take priority and mention has been made to over-staffing, and it is not uncommon for 80 + % of extension budgets going on salaries. One consequence is that the necessary recurrent support costs to cover transport, travel allowances, fuel, extension support material are severely restricted.

Lending agencies are reticent regarding lending to cover recurrent costs, unless it is crucial for a project; this is not necessarily case with capital costs. It is still rare to find an adequately capitalised extension service with sufficient housing, vehicles and equipment.

The lack of resources are not just confined to the extension institutions. If development is to succeed, it will dependent equally on the supply of essential inputs, for example, credit, breeding stock, drugs etc. If production is to increase, it is essential that both a market and a marketing infrastructure are available to handle surplus production. This section has not intended to be deliberating pessimistic, its aim has been to highlight some of the main constraints, so as to assist with a examination of potential strategies.

Potential Extension Strategies to Promote Livestock Production

Some premise need to be stated from the onset:

- that the resources allocated to livestock extension are limited and likely to remain so, therefore, optimal use must be made those resources available,

- that a mandate for the livestock extension will remain one of many retained by the general extension worker and that scope for further livestock extension activities is probably limited, and

- whatever strategies are adopted, regardless of the approach or system used, they must be suited to local conditions; flexible enough to be able to respond to changing circumstances; incorporate producer participation at all levels in order to ensure sustainability in both in the long and short-term.

National Planning

Clearly, there has to be policies to provide the structure and guidance to the livestock sector. These policies should be realistic as well as being economically and environmentally sound and, hopefully, would have taken into account the existing (or pipeline) situation regarding: institutional, infrastructural and marketing aspects. To be realistic, policies must take account of both the aspirations and capability of the producer.

Policies will differ both between and within countries and may be commodity or sectorially orientated. Policies should provide a framework, but must also be accompanied by clearly stated and attainable objectives. Again, objectives will need to be determined at the regional and local levels.

Strategies, of which extension is just one, need to be determined that can effectly achieve these objectives. Such strategies could include, for example, strengthening research, training more specialists, investment in marketing and infrastructure. And, of course, the question must also be asked whether such strategies can economically achieve their objectives - this would require some degree of for cost-benefit analysis.

We believe that there is an important role for both the major lending agencies and the international technical organisations, such as FAO, in providing assistance to governments in determining policies, objectives and priorities at the sectorial level.

Setting Priorities

Given that resources are likely to continue to be limited, it will be essential that efforts to develop livestock production are prioritized and targeted within the broader policy framework.

This raises the question of which sectors should be targeted and where would be the appropriate entry point. Efforts should be targeted to those situations where interventions are likely to have the greatest impact or benefit. Livestock production systems range along a continuum of sophistication that starts with the scavenging or extensive, minimal input and limited output systems, through semi-commercial to sophisticated, commercial production.

The majority of livestock are held in the small-holder sector, small but widespread increases in productivity could, overall, make a considerable impact. The problem is that in risk-adverse communities where animals are kept for many purposes, the minimal input some output systems take some beating. To change such systems is not easy and we suggest should not be an objective on a blanket approach basis.

Improvements are, however, possible as has been shown by FAO's small ruminant projects in Togo and The Gambia where appropriate technologies, based on improved husbandry and basic inputs, have been introduced and accepted along with corresponding increases in productivity. However, they are expensive in that they required intensive extension inputs at levels, which we would suggest, are not sustainable on a wider basis. What such extension exercises do achieve, however provided they are successful, is to activate producers to benefits of improved animal husbandry, and move them further along the continuum.

It is not just extension that can provide the impetus for change, other circumstances can also be responsible. Nigeria provides a good example, where sound technologies based on browse trees integrated into the farming systems along with animal health interventions (developed by the ILCA Small Ruminant Programme) have had a slow uptake in western Nigeria, where land pressures are such that farming practices are not under sufficient pressure to change. In contrast, in eastern Nigeria with higher human populations, land use is under greater pressure. Virtually overnight in many villages local by-laws were enacted which confined (previously free ranging) small ruminants, suddenly animals had to be managed and inputs provided. Gradually, some farmers gave up keeping sheep and goats, whereas other started to specialised. Here was situation that was amenable to change and would have been a suitable area in which to focus extension effort.

Where effective producers associations (formal or informal) are operational they should provide an appropriate focal point for an extension programme. If not, the formation of such groups should be a priority.

In targeting areas or sectors, attention must be given to ensure that the necessary support and ancillary services, input supply and distribution, and marketing infrastructure are available. We would suggest that in this regard there is much to be said for targeting livestock extension programmes within the overall umbrella of an on-going integrated agricultural development programme.

Lastly, but not least, there has to be a willingness on behalf of the farming community to participate in a programme. Ideally, the initiative would have originated from the community, not as all too often happens, an programme being sold to the producer.

Problem Identification and Potential Solutions

Having selected target areas/sectors it is necessary to identify problems and constraints, in particular those that can be readily rectified. Rapid Rural Appraisal (RRA) techniques provide a convenient methodology for this purpose. The RRA should contain a Subject Matter Specialist Knowledgeable not only of the principal production systems, but also conversant with the cultural and socioeconomic conditions of the area. The role of the Subject Matter Specialist is crucial, without adequate training and practical experience, there is a real danger of inappropriate technologies being promoted.

Proposed interventions need to be practical, acceptable and stand a realistic chance of achieving their objectives. It is important that other aspects, apart from pure technical issues, are clearly understood, for example: labour availability, reasons for keeping animals, women involvement, cash resources and markets are fully taken into account. If necessary On-Farm-Testing (OFT) should be carried out as part of the participatory process involving research, extension and the producer.

Detailed Workplans

Immediate objectives need to be prepared with detailed workplans. The workplan should detail manpower requirements, responsibilities and, at least at a rudimentary level, a critical path analysis to identify crucial points in the programme.

The impact of the extension programme can be evaluated by a) regular RRAs and b) regular monitoring of animal productivity from selected producers. Simple recording of herd/flock dynamics (births, deaths, sales, purchases), estimates of milk yields, if applicable, and/or size/weight estimates, combined with local market information can produce valuable information of actual performance.

Extension Methodologies

The efficient use of extension technologies is not new. Interpersonal techniques are effective, providing the agent has credibility and a good product to sell, but are particularly expensive in terms of staff and resources. The approach is justifiable in establishing a network of contact families who we believe should be the principle agents of change. Women have a prominent role in small-scale animal production systems, but, in many countries it is difficult to deal directly with female household members for cultural reasons. Furthermore, we still have a long way to go in recruiting sufficient female extension agents. The concept of the contact family is that an agent deals with both the husband and wife who, in turn, disseminate information to their respective peer and gender groups. The extension process, thereafter, evolves around the contact family who act as demonstrators, risk takers and sources of information. Demonstrations, field days, group meetings are the techniques used with which we are all familiar. This is the classic "extension by example" approach.

General awareness and continued support for such programmes are provided through the mass media: radio, television, newsletters etc. Certainly, we believe that greater use could be made of these techniques, especially as both the necessary hardware and software is nearly always available. What is so often lacking is the adequate skills to use them to greatest effect.

Extension Strategies

Given that appropriate policies are available, priorities have been established, appropriate technologies are tested and available, workplans and objectives have been set - there remains the final hurdle of how to realistically implement such a programme.

Training cum Development: This is an approach that FAO (AGA) is developing in some its projects to facilitate livestock extension and development. The concept is to use a small, multi-disciplinary, task force within a selected area, to promote livestock production through a combination of training and development activities. It recognises a number of important points:

- that initially livestock extension must be channelled through the existing extension services,
- that general extension workers have little training or understanding of basic concepts of animal husbandry,
- that the extension worker has other duties and that limited time will be devoted to livestock extension, and
- that it is unusual to find appropriate technical packages either available or it use.

The approach adopted has been to follow through a series of sequential steps.

- identification of target areas with the relevant authorities,
- establishment of a local task force consisting of SMSs, researchers, extensionists and trainers who will be responsible for implementing the programme,
- preparation of appropriate technical and extension messages,
- undertaking of a training needs assessment of extension staff,
- preparation of the necessary extension support materials,
- initial training of participating extension staff covering basic concepts to provide a framework on which to hang further skills and knowledge,
- followed by further progressive training only in those areas of knowledge and skills directly relevant to the extension message and combined with supervised development activities being undertaken between training sessions, and
- monitoring and evaluation.

A number of lessons have been learnt but the approach still needs further development and modification. In one particular project, problems arose in a number of areas, notably with regard to the involvement of a) producers in the both planning and training stages b) the support of the extension worker's immediate supervisions and c) the lack of initial an rapid rural appraisal. Out of necessity, training was undertaken at training centres and we believe that residential courses at training centres are not necessarily conductive to good training - invariably they have neither the facilities or the trained staff.

With dutch bilateral assistance the International Poultry Centre in Indonesia has taken its extension staff and farmer training programme out of the Centre. Training is now being undertaken with small training teams in selected districts, which travel around and hold training sessions for both farmers and extension staff in the villages, using village facilities. We believe this to be an extremely promising exercise that should be followed closely.

What is clear, regardless of the extension methodology adopted, is the crucial role played by the Subject Matter Specialist. It is at this level that greater emphasis is required to ensure that they are adequately trained and have the opportunity to gain necessary and relevant practical experience.

Sustainable Approaches for the Future. All the alternative approaches mentioned are viable and sustainable in the short-term, in that they prioritize and optimise existing resources. However, we would suggest that government funded livestock extension, as we know it, is not sustainable in the long term in developing countries.

Greater effort needs to be given to examining ways in which the advisory role can be transferred to respected livestock producers within the community. Not only would this have considerable cost savings but would almost certainly be more effective than the use of inexperienced extension agents who lack in credibility. We believe that the development of producer associations or co-operatives are likely to have a pivotal role in the provision of livestock extension services in the future.

Externally assisted projects should examine ways in which the infrastructure, capital or credit can be provided to establish services (advice and inputs) that farmers would be willing to pay for. For example, if we take a hypothetical case of a mixed farming area where cereals and grain legumes are grown and livestock are an important component of the farm system. Straw, plus grain supplements, are likely to be a major components of the diet. A small unit providing the services of say a hammer mill, simple mixer and maybe a straw chopper, with payment being made in cash or kind, would be a welcomed and utilised service. Supplementation with molasses and urea would have an important beneficial impact on the nutrition and, consequently, productivity of the animals. However, provision of the inputs is often restricted by the logistics of distribution. A bulk storage tank for molasses and a store for urea, linked into a wider distribution network, would make such ingredients readily available. One could envisage the situation where the operation of such a unit could be leased to a respected livestock producer in return for a commitment to undertake basic training and advise other farmers. Extra duties could be provided on a contract basis, for example, tagging and recording animals using AI, operating bull centres or maintaining demonstration plots. Such a system could easily be developed under the umbrella of an active producer association.

We strongly believe that there remains scope for more imaginative approaches to be developed to provide more effective livestock extension in developing countries.

As producers associations develop, which must be a major objective of an extension service, they will reach the stage where they are able to provide their own advisory services along with input supply and marketing services. We believe that the experience provided by the Operation Flood programme in India is an example of the potential of producer associations and provides valuable lessons for the future.

Farming Systems Methods in the Planning, Implementation and Monitoring of Sustainable Livestock Development

Farming Systems Research (FSR) is an approach that (a) takes a holistic view of the whole farm as a system, (b) focuses on the relationships between the various components under the control of the farm household and of the interactions of these components with physical, biological and socioeconomic factors under the household's control, and (c) aims at enhancing the efficiency of farming systems by focusing agricultural research to generate and test improved technologies. Basic characteristics of the approach is that it focuses on small farmers and is holistic, integrated, location specific and dynamic. In the case of livestock the results may be applicable across a wider range of situations. Historically, the concept was developed with a arable bias and it is only in the last decade that the livestock component has been added. Several Asian countries including the Philippines, Indonesia, Pakistan, India and Sri Lanka are now employing FSR methods in livestock development projects.

Initially, livestock interactions were only considered important in mixed farming systems where feed and labour interactions where obvious and easy to quantify. Pastoral and specialised range systems still have not yet been extensively subjected to the FSR approach. International Centres have taken a lead in testing the systems approach to farm level development and there is now increased interest to incorporate such methodologies into national programmes. International Development Organisations (IDOs), such as, FAO and the World Bank, are now encouraging the FSR approach in dealing with the broader issues of sustainable agriculture development; the environment; women in development and structural reform.

The current trend is towards more specialised, high input-output production systems based on modern technology, socioeconomic factors and supported, as necessary, by instruments of policy, such as, subsidies, tariffs and quotas. Farming systems, particularly those producing grain and oilseed crops depend on crop rotations and other diversification strategies and are contrary to the trend towards specialisation and intensification of most agricultural operations. Diversification helps reduce risk by spreading it amongst a number of crop and animal activities. The most common diversification strategy is the combination of crop and livestock enterprises, and many grain legumes found in the crop rotations provide valuable crop residues as well as a valuable source of nitrogenous fertilizer. Similarly manure recycling provides a valuable source of fertilizer and animals may provide the primary means of traction. Livestock development projects concerned with mixed farm production systems will certainly benefit from the FSR perspective.

Livestock Development Projects

Great concern has been expressed over the performance of livestock development compared to the other agricultural sectors. (FAO, 1990) and the experience regarding livestock projects is disturbing. Such projects often fail to deliver tangible products to the beneficiaries and seldom trigger variables that can lead to long term growth. A recent Asian Development Bank review (ADB, 1990) found deficiencies and lessons learned from Bank financed projects. These included:

Pre-feasibility Planning Stage:

- an absence of reliable basic data covering animal populations, production parameters, demand and consumption estimates, marketing channels, commodity prices and income and cross elasticities, and
- that project participants were not clearly identified.

Project Implementation Stage:

- that a lack of counterpart funding was a major constraint to project implementation, and
- shortages in manpower and local commitment.

Project Evaluation Stage:

- that more stress is given to the physical completion of infrastructure and distribution of animals than is given to less quantifiable changes in labour utilisation, employment creation, use of fodder and grasses, improved human nutrition, and the long term genetic improvement of livestock.
- benefits from livestock projects are usually derived over a longer time frame than the normal disbursement period, a factor that is not taken into account when projects are evaluated, usually at the end of the disbursement period, and
- this has meant that livestock projects often compare poorly with investments in other agricultural sectors.

The Asian Development Bank experience indicates that there is considerable scope for modifying the approach to livestock development projects. A prerequisite is to obtain accurate and relevant information, especially at the farm level, on which to plan projects. Equally important is to ensure the direct participation of the beneficiaries in the planning process. The planning, implementation and monitoring of the proposed activities requires multi-disciplinary teams experienced in the describing the existing situation, identifying constraints and designing, testing, evaluating and extension of appropriate technology. There is also a concern regarding the ability of many professionals (veterinarians and animal husbandry specialists) to adequately deal with non-commercial aspects of animal production.

Many training establishments, including those in developing countries, are not producing graduates with the necessary skills and understanding required to work with smallholder farm systems. In particular, small stock, such as rabbits, are poorly represented and there is a lack practical training which is important to gain the confidence of producers.

The Role FSR Methods in Planning and Implementation of Livestock Projects

Planning Stage

FSR data collection techniques employ both formal and informal approaches. The Rapid Rural Appraisal technique and informal diagnostic surveys have shown merit in generating valuable farm level data relevant for project design. These survey approaches are quick, cost efficient and can generate both quantitative and qualitative data. These techniques require assistance from a range of expertise and, as such, the demand on skilled manpower is high. Only essential information necessary to define project boundaries, identify constraints and opportunities, develop basic technical parameters is generated. Several surveys conducted in the Baluchistan province of Pakistan by the Arid Zone Research Institute in collaboration with ICARDA (Syria) proved useful in planning a recent IFAD funded Integrated Range and Livestock Development Project.

In Pakistan, FSR generated information has proved invaluable and is used by donor missions and government agencies to assist in conceptualising projects and as a source baseline data. Amir and Ahmed (1990) cite several cases from Pakistan where FSR micro-data has used by policy makers particularly in Planning and Development Departments who are responsible for project screening.

Diagnosis Stage

Detailed case studies, farmer meetings and diagnostic surveys have provided valuable information that has identified problems and constraints related to animal health and productivity, social economics and the role of women. In particular, it is capable of identifying farmers aspirations. (FSSP, 1985 and 1987). Rapid appraisals of rural markets can generate basic information regarding marketing practices, margins, transportation costs, problems related to input supply regarding livestock (Young and Amir, 1988).

There is obvious scope for further donor missions and government agencies to involve FSR researchers, extension workers and farmers to identify both farm level and institutional constraints and to provide field level inputs to define appropriate approaches and potential solutions.

In Indonesia, the Directorate General of Livestock Services and university livestock faculties have initiated an on-farm livestock project that is coordinated under the Upland Agriculture Conservation Project. By using FSR diagnostic tools, livestock technologies can be disseminated and monitored in an effective manner.

Technology Testing

Where new technologies are being introduced; testing, refining and tailoring them to meet farmers needs is an important stage (Zandstra, 1985). Some livestock projects assume that commercially available techniques are equally applicable under smallholder conditions and often this assumption leads to the failure of many livestock projects. For example, imported exotic breeds seldom perform as expected under smallholder management conditions. Preliminary testing using FSR methods assist in ensuring that new technologies are appropriate, and it may be advantageous to establish on-farm evaluation units within projects. This project component should be staffed with a multi-disciplinary team to ensure the smooth transfer of technology. At this stage it will be necessary to that the following into account:

- economic viability,
- technical feasibility,
- social soundness,
- political acceptability, and
- institutional capacity.

These prerequisites are crucial for the successful transfer of improved technologies. Livestock projects that have a wide array of technology options will benefit more from on-farm testing, compared to situations where totally new solutions have to be devised. In some cases, technologies practised by progressive farmers, when properly extended, may find greater acceptance than imported technology that is unfamiliar under village situations. On-farm testing and evaluation can help ensure that inappropriate technologies are not introduced by projects. An example of such inappropriate technologies is found in the introduction of Holstein-Friesian cows in Indonesia. Lack of milking machines, high feed requirements and disease susceptibility made these cows a liability to many farmers who had obtained them on loans. Projects that focus on animal distribution programs, especially of exotic stock, should be aware of the limitations of such an approach to small farm development.

Bibliography

Ahlgren, G.H.: *Forage Crops,* McGraw-Hill, New York, 1956.

Astley Maberly, C. T.: *Animals of East Africa,* Hodder & Stoughton, London, 1966.

Boehrer, Bruce: *A Cultural History of Animals in the Renaissance,* Oxford, UK: Berg Publishers, 2007.

Bogdan, A.V.: *Tropical Pasture and Fodder Legumes,* Longmans, London, 1977.

Bright, M.: *Animal Language,* BBC Publications, London, 1984.

Carl Cohen and Tom Regan: *The Animal Rights Debate,* Rowman & Littlefield, Lanham, MD, 2001.

Cheeke, P. R.: *Applied Animal Nutrition: Feeds and Feeding,* McGraw-Hill, New York, 1998

Church, D. C.: *Livestock Feeds and Feeding,* McGraw-Hill, New York, 1997

Cloudsley-Thompson, J. L.: *The Zoology of Tropical Africa,* W. W. Norton, New York, 1969.

Cullison, A. and R. S. Lowrey, *Feeds and Feeding,* McGraw-Hill, New York, 1986

Curtis, H. and N. S. Barnes: *Biology,* Worth Publishers, Inc. New York, 1989.

Davidson, Peter Hobley. George Orwell: *A Literary Life,* St. Martin's Press, New York, 1996.

Dolins, Francine: *Attitudes to Animals: Views on Animal Welfare,* Cambridge University Press, Cambridge, 1999.

Ensminger, J. E. Oldfield, and W. W. Heinemann, *Feeds and Nutrition,* McGraw-Hill, New York, 1998

Ensminger, R. M.: *The Stockman's Handbook,* McGraw-Hill, New York, 1992

Estes, Richard Despard: *The Behaviour Guide to African Animals,* University of California Press, Berkeley, CA, 1991.

Fairey, D.T., & Hampton, J.G.: *Forage Seed Production of Temperate Species,* CAB, Farnham Royal, 1997.

Fogle, Bruce: *Pets and Their People,* The Viking Press, New York, 1983.

Fox, Michael Allen: *The Case for Animal Experimentation,* University of California Press, Berkeley, CA, 1986.

Frame, J., Charlton, J.F.L., & Laidlaw, A.S.: *Temperate Forage Legumes*, CAB Interational, Wallingford, 1998.

Frame, J.: *Improved Grassland Management*, Farming Press, Ipswich, 1992.

Frederick, Zeuner E.: *A History of Domesticated Animals*, Hutchinson, London, 1963.

Friend, Tim: *Animal Talk: Breaking the Codes of Animal Language*, Free Press, New York, 2004.

Fudge, Erica: *Brutal Reasoning: Animals, Rationality and Humanity in Early Modern England*, Cornell University Press, Ithaca, 2006.

Gallistel, C.R.: *Animal Cognition*, MIT Press, Cambridge, 1992.

Gates, P.: *Animal Communication*, Cambridge University Press, Cambridge, 1997.

Giorgio, Agamben: *The Open: Man and Animal*, Stanford University Press, UK, 2004.

Hacker, J.B.: *Nutritional Limits to Animal Production from Pasture*, CAB, Farnham Royal, 1981.

Hanson, A.A., Barnes, D.K., & Hill, R.R.: *Alfalfa and Alfalfa Management*, Wisconsin, Madison, 1988.

Hanson, C.H. : *Alfalfa Science and Technology*, Madison, Wisconsin, 1998.

Harris, Marvin: *The Sacred Cow and the Abominable Pig: Riddles of Food and Culture*, Touchstone Books, New York, 1987.

Hearne, Vicki: *Animal Happiness*, HarperCollins, New York, 1994.

Hitchcock, A.S., & Chase, A.: *Manual of Grasses of the United States*, USDA, 1971.

Holmes, W.: *Grass, its Production and Utilisation*, Blackwell, Oxford & London, 1989.

Iwago, Mitsuaki: *Serengeti: Natural Order on the African Plain*, Chronicle Books,San Francisco, CA, 1987.

Jones, M.B. & Lazenby, A.: *The Grass Crop,* Chapman & Hall, London, 1988.

Julian Baldick: *Animals and Shaman: Ancient Religions of Central Asia*, New York University Press, New York, 2000.

Juliet: *Domesticated Animals from Early Times*, University of Texas Press, Austin, TX, 1981.

Karen, Allen Miller: *The Human-Animal Bond: An Annotated Bibliography*, Scarecrow Press, Metuchen, NJ, 1985.

Kevin Dolan: *Ethics, Animals, and Science*, Blackwell Science, Malden, MA, 1999.

Kiss, Agnes: *Living with Wildlife: Wildlife Resource Management with Local Participation in Africa*, World Bank, Washington, D.C., c1990.

Langley , Gill: *Animal Experimentation: The Consensus Changes*, Chapman and Hall, New York, 1989.

Lawick, Hugo van: *Among Predators and Prey*, Sierra Club Books, San Francisco, 1986.

Lu, F.C. & Rendel, J.: *Anabolic Agents in Animal Production*, FAO/WHO Symposium, Rome, 1975.

Mason, Jim and Peter Singer: *Animal Factories*, Crown Publishers, New York, 1980.

Masson, Jeffrey Moussaieff: *Dogs Never Lie About Love: Reflections on the Emotional World of Dogs*, Crown Publishers, New York, 1997.

McDonald, P., Edwards, R.A., Grenhalgh, J.F.D., & Morgan, C.A.: *Animal Nutrition*, Longmans, London, 1995.

Morton, Eugene S.: *Animal Talk: Science and the Voices of Nature,* Random House, New York, 1992.

Natz, D.: *Feed Additive Compendium*, McGraw-Hill, New York, 1977

O'Neill, Terry: *Readings on Animal Farm*, Greenhaven Press, San Diego, 1998.

Orwell, George: *Animal Farm: A Fairy Story*, Signet Classics, Orlando, 1996.

Pietro Croce: *Vivisection or Science?: An Investigation into Testing Drugs and Safeguarding Health*, Zed Books, New York, 1999.

Pond, W. G.: *Basic Animal Nutrition and Feeding*, McGraw-Hill, New York,, 1995

Randall, D., W. Burggren, and K. French: *Eckert Animal Physiology*, W.H. Freeman and Company, New York, 1997.

Raymond, F., Redman, P., & Waltham, R.: *Forage conservation and Feeding*, Farming Press, Ipswich, 1986.

Roger, J.: *Buffon: A Life in Natural History,* Cornell University Press, Ithaca , NY, 1997.

Roy, J.H.B.: *Studies in the Agricultural and Food Sciences: the Calf*, Butterworths, London, 1980.

Ruth Ellen Bulger: *The Ethical Dimensions of the Biological and Health Sciences*, Cambridge University Press, New York, 2002.

Service, Robert: *A History of Modern Russia*, Harvard University Press, Cambridge, Mass, 2005.

Sharpe, Robert: *Science on Trial: The Human Cost of Animal Experiments*, Awareness Books, Sheffield, UK, 1994.

Skerman, P.J., & Riveros, F.: *Tropical Grasses*, FAO, Rome, 1989.

Skerman, P.J., Cameron, D.G., & Riveros, F.: *Tropical Forage Legumes*, FAO, Rome, 1988

Snaydon, R.W.: *Managed Grasslands,* Elsevier, Amsterdam & London, 1987.

Steve Baker: *The Postmodern Animal,* Reaktion Books, London, 2000.

Stuart, Chris and Tilde Stuart: *Africa's Vanishing Wildlife,* Smithsonian Institution Press, Washington, D.C., 1996.

Taylor, R.E. and R. Bogart: *Scientific Farm Animal Production: An Introduction to Animal Science,* MacMillan, New York, 1988.

Ted Benton: *Natural Relations: Ecology, Animal Rights and Social Justice,* Verso, London, 1993.

Umali, D. & Schwartz, L.: *Public and Private Agricultural Extension: Beyond Traditional Frontiers,* Washington, DC, World Bank, 1994.

VanderWal, P.: *Anabolic Agents in Animal Production,* Environmental Quality and Safety, Suppl. 1976.

Walton, P.D.: *Production and Management of Cultivated Fodders,* Reston Publishing, Reston, VA, 1982.

Whyte, R.O, Moir, T.R.G., & Cooper, J.P.: *Grasses in Agriculture,* FAO, Rome, 1959.

Whyte, R.O., Nillson-Leissner, G., & Trumble, H.C.: *Legumes in Agriculture,* FAO, Rome, 1953.

Wieczynski, Joseph L.: *The Modern Encyclopedia of Russian and Soviet History,* Academic International Press, Gulf Breeze, Fla, 1976.

Willis, R.G.: *Signifying Animals: Human Meaning in the Natural World,* Unwin Hyman, London, 1990.

Yarri, Donna: *The Ethics of Animal Experimentation: A Critical Analysis And Constructive Christian Proposal,* Oxford University Press, Oxford, 2005.

Index

□□□